T0282617

CAMBRIDGE GEOLOGICAL SERIES

A MANUAL OF SEISMOLOGY

A MANUAL
OF
SEISMOLOGY

BY

CHARLES DAVISON, Sc.D.

CAMBRIDGE
AT THE UNIVERSITY PRESS
1921

CAMBRIDGE
UNIVERSITY PRESS

University Printing House, Cambridge CB2 8BS, United Kingdom

Published in the United States of America by Cambridge University Press, New York

Cambridge University Press is part of the University of Cambridge.

It furthers the University's mission by disseminating knowledge in the pursuit of education, learning and research at the highest international levels of excellence.

www.cambridge.org
Information on this title: www.cambridge.org/9781107690202

First published 1921
First paperback edition 2014

A catalogue record for this publication is available from the British Library

ISBN 978-1-107-69020-2 Paperback

PREFACE

My chief difficulty in writing this textbook has been to decide on what should be included and what excluded, to determine in fact the relative importance of different branches of a subject which is of growth so modern that its treatment has not yet become defined. Until near the close of the last century, seismology was regarded as a department of geology, but, while its growth in that direction has by no means ceased, the more recent advances have been largely the work of mathematicians and physicists. As this volume belongs to a series of geological manuals, much of that work is here left unnoticed, and readers who wish to know more of these recent developments may be referred to Prof. Knott's *Physics of Earthquake Phenomena* (Oxford University Press) and Dr G. W. Walker's *Modern Seismology* (Longmans).

Partly in order to save room, I have omitted several subjects which are usually included within the domain of seismology. Some of these (such as the bending of the earth's crust by tidal loading and the occurrence of many micro-tremors) have in reality no connexion with earthquakes. Others (such as the effects of earthquakes on men and animals, the rotation of columns and experiments on the velocity of earth-waves) are of interest but are not essential to the subject.

Though a few historical notes are inserted here and there, no attempt is made to provide a history of seismology. My aim has been to give an outline of our present knowledge, which of course is usually in advance of the work of pioneers. The number of references to any writer in the index is thus rather a testimony to the recency of his achievements than to the actual part which he has taken in promoting the science. Otherwise, there would have been many more references to the work of Perrey, Mallet and Milne. Readers who desire further information on the history of the science may consult a series of papers on the "Founders of Seismology" in the *Geological Magazine* for 1921.

With regard to references generally, it may be convenient if I mention here the plan which I have followed. At the beginning of nearly every chapter, I have given a list of the memoirs which the reader who desires fuller information may study with advantage. Throughout the chapter, these are quoted under the author's name. References required in each section are collected in a footnote at the end of the section, and, whenever a choice is available, I have preferred those to English or easily accessible works. The abbreviated titles of the more important books and journals are given after the Table of Contents. For two of the diagrams (Figs. 11 and 21), I am indebted to the courtesy of the Directors of the Cambridge Scientific Instrument Company.

CHARLES DAVISON.

CAMBRIDGE.
March, 1921.

CONTENTS

LIST OF ILLUSTRATIONS

*available for download from www.cambridge.org/9781107690202

ABBREVIATIONS

In addition to the usual abbreviations for well-known scientific journals, the following are used in the footnote references:

Boll. Soc. Sis. Ital. Bollettino della Società Sismologica Italiana.

Bull. Eq. Inv. Com. Bulletin of the Imperial Earthquake Investigation Committee (Tokyo).

Bull. Seis. Soc. Amer. Bulletin of the Seismological Society of America.

Dutton. C. E. Dutton, The Charleston Earthquake of August 31st, 1886. Ninth Annual Report, U.S. Geological Survey, 1889, pp. 209–528.

Lawson. The Californian Earthquake of April 18, 1906 (edited by A. C. Lawson), vol. 1 and atlas, 1908; vol. 2 (by H. F. Reid), 1910.

Oldham. R. D. Oldham, Report on the Great Earthquake of 12th June, 1897. Mem. of the Geol. Surv. of India, vol. 29, 1899, pp. 1–379.

Publ. Eq. Inv. Com. Publications of the Imperial Earthquake Investigation Committee in Foreign Languages (Tokyo).

Seis. Journ. Seismological Journal of Japan.

Trans. Seis. Soc. Japan. Transactions of the Seismological Society of Japan.

CHAPTER I

INTRODUCTION

1. An *earthquake*, in its widest sense, is a result of any sudden displacement of the earth's crust, either on or beneath its surface. The causes of the displacement may be natural or artificial, or partly natural and partly artificial, in their origin. The term is usually restricted, however, to movements of natural origin and to those which take place below the earth's surface; and it is in this sense that it will be used in this volume. Thus, the tremors caused by wind or sea-waves, by landslips or rockfalls, will not be regarded as true earthquakes; while those associated with volcanic operations or with the growth or shaping of the earth's crust will form the chief subjects of our inquiry.

A disturbance of this nature is said to be *seismic*, and the science that deals with the phenomena and origin of such earthquakes is called *seismology*.

An earthquake felt at sea, and therefore consisting of movements propagated through the sea, is called a *seaquake*.

The initial displacement which results in an earthquake usually occurs at some depth below the surface, it may be of one or more miles. The displacement generates series of waves, which spread outwards with great velocity in all directions. As the waves pass any particle of rock, they cause that particle to move rapidly to and fro, and it is mainly this vibratory movement which produces the sensation of an earthquake. In some rare cases, however, the initial displacement, though in part deep-seated, is continued up to the surface, and the vibratory motion is then complicated by the mass-displacement of the crust.

2. Each complete to-and-fro movement of an earth-particle is called a *vibration*.

Let us suppose that a particle is executing vibrations along a straight line at right angles to AB (Fig. 1), and that it can trace its course on a sheet of paper placed below it. If the paper at the same time be moved with uniform velocity in the direction *BA*, the path traced by the moving particle will take some form like that of the curve *APCQB*. In the simplest case, when the vibrations are not large, the motion is that known as simple harmonic motion, and the curve so drawn during one complete vibration is that corresponding to one complete period of the

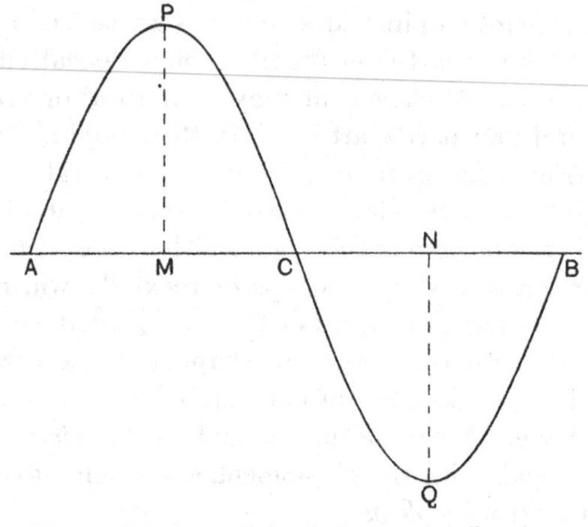

Fig. 1. Diagram of simple harmonic vibration.

curve of sines. The portion *APC* or *CQB* corresponds to a semi-vibration. The time taken to execute a complete vibration (represented by the line *AB*) is called the *period* of the vibration. The distance between the extreme positions, *P* and *Q*, of the particle measured along a line perpendicular to *AB* (that is, the sum of the distances *PM* and *NQ*) is called the *range* of the vibration. Half the range (that is, *PM* or *NQ*) is called the *amplitude* of the vibration.

The *velocity* of the particle is the rate at which it is changing its distance from some fixed point in its line of motion, say, from its position of rest in *AB*. The *acceleration* of the particle is the rate at which its velocity is changing*.

* If the displacement at time t be $a \sin(\lambda t + \epsilon)$, where a, λ and ϵ are constants, the velocity and acceleration at the same moment are

$$a\lambda \sin\left(\lambda t + \epsilon + \frac{\pi}{2}\right) \text{ and } a\lambda^2 \sin(\lambda t + \epsilon + \pi).$$

Thus, the amplitude or maximum displacement is a, the maximum velocity

The angle which the path of a vibrating particle at the surface of the earth makes with the horizontal plane through it is the *angle of emergence*.

3. While the term *shock* is often used as synonymous with earthquake, it should, strictly speaking, be confined to those vibrations which are sensible to the body of an observer.

The vibrations which are so rapid that they affect the ear of an observer constitute the *earthquake-sound*. When the sound is heard without any attendant shock, it is called an *earth-sound*.

The terms shock and sound are convenient, though there is no definite boundary between the two sensations, for the vibrations which produce the lowest sounds give rise to a quivering that is sensible to other parts of the body than the ear.

Vibrations may be insensible to human beings either from the smallness of their amplitude or the lengths of their periods. Vibrations of the former class give rise to *earth-tremors* or *micro-tremors*, those of the latter class to *earth-pulsations*.

In French works, the following terms are sometimes used: *seisms* for earthquakes generally, *microseisms* for earth-tremors, *macroseisms* for sensible earthquakes, *megaseisms* for destructive earthquakes, and *teleseisms* for earthquakes of distant origin. In English, such terms are uncouth, but the corresponding adjectives are occasionally useful.

4. The place or region within which an earthquake originates is usually called (after Mallet) the *seismic focus* or *focus*, occasionally the *centre*, *centrum* or *hypocentre*. The chief objection to all such terms is the implication that the region in question is a point. Less objectionable, perhaps, is the word *origin*, meaning place of origin. No term, however, has obtained such currency as seismic focus or focus, and one or other will be used for the future, it being understood that the focus is a region often of great extent.

is $a\lambda$, and the maximum acceleration $a\lambda^2$. If T be the period of a complete vibration, we have $\lambda T = 2\pi$ or $\lambda = 2\pi/T$. Thus, the maximum velocity is $2\pi a/T$, and the maximum acceleration $4\pi^2 a/T^2$. It is shown in text-books of dynamics, or it may be deduced from the above expressions, that the maximum velocity is attained when the particle is passing through its position of rest (as at A, C or B), and the maximum acceleration when the particle is farthest from its position of rest (as at P or Q).

The area on the earth's surface vertically above the seismic focus—that is, the projection of the focus on the surface—is called the *epicentre*. The same objection applies to this term as to the words seismic focus and hypocentre. The epicentre is not a point, but an area of some, often of considerable, extent in at least one direction.

5. The *intensity* of an earthquake is proportional to the maximum acceleration of its vibrations, which is usually measured in millimetres per second per second.

An *isoseismal line*, or simply an *isoseismal*, is a line drawn through all places at which the intensity of a shock is the same. The *meizoseismal area* of an earthquake is the area within which the intensity is greatest. The term is somewhat indefinite and should perhaps be confined to the area included within the innermost isoseismal line*.

The *disturbed area* of an earthquake is the district within which the shock is perceptible to the unaided senses.

An *isacoustic line* is a line drawn through all places at which the same percentage of the total number of observers in it are capable of hearing the earthquake-sound.

The *sound-area* of an earthquake is the district within which the earthquake-sound is audible to some observers without instrumental aid.

6. When the epicentre of a great earthquake is submarine, the earthquake may be followed by a series of sea-waves. These are known as the *seismic sea-waves*, in Japan as *tsunamis*. In popular writings, they are frequently called "tidal waves," an obvious misnomer, for such waves, though simulating tides of short period, are of entirely different origin.

7. A strong earthquake is sometimes, though not always, preceded by a small number of slight shocks, and is invariably followed by a large number of a similar character.

The great earthquake, with reference to the others, is called the *principal shock* or *principal earthquake*. The minor shocks are known as *accessory shocks*, those which occur before the

* Mallet, to whom we are indebted for several of our terms, also defines a *coseismal line* as a line drawn through all points at which the same phase of the movement is felt at the same instant. The term is now seldom used.

principal earthquake as *fore-shocks*, and those which occur after it as *after-shocks*.

In districts near that in which an earthquake originates, slight shocks may be precipitated by the changes of stress introduced by the occurrence of the earthquake. These are known as *sympathetic earthquakes*.

8. Various classifications of earthquakes have been proposed and suitable names devised for them (see sect. 34). For the present, it will be sufficient to refer to the two classes of volcanic and tectonic earthquakes. *Volcanic* earthquakes are those which precede, accompany, or follow, the operations of a volcanic eruption or are due to displacements within the mass of a volcano. *Tectonic* earthquakes are the results of the growth of the earth's crust, that is, of the deformations to which the form of its surface-features is ultimately due*.

* The present volume is concerned with the phenomena of earthquakes in general, and the description of individual earthquakes lies beyond its range. It is important, however, that such descriptions should be studied. The most complete reports of recent earthquakes are those by C. E. Dutton on the Charleston earthquake of 1886 (*Ann. Rep. U.S. Geol. Surv.*, vol. 9, 1889, pp. 209–528), R. D. Oldham on the Assam earthquake of 1897 (*Mem. Geol. Surv. India*, vol. 29, 1899, pp. 1–379), and A. C. Lawson (editor) on the Californian earthquake of 1906 (*Report of the State Earthquake Investigation Commission*, vol. 1 and atlas, 1908; vol. 2, 1910). Frequent references to these valuable reports will be found in the following pages, the names of the authors and editor being given without the full titles of the books. Brief descriptions of various earthquakes, from the Neapolitan earthquake of 1857 to the Assam earthquake of 1897, are given in C. Davison's *Study of Recent Earthquakes* (Contemporary Science Series), 1905.

CHAPTER II

SEISMOGRAPHS

9. Of the instruments which have been designed for recording earthquakes, *seismoscopes* are intended merely to register the occurrence or the time of occurrence of an earthquake and they will not be considered in this chapter. The object of *seismometers*, *seismographs* or *tromometers* is to record in detail at every moment of an earthquake the position of the ground relatively to its position of rest, so that, from the record or *seïsmogram*, we may determine the amplitude, direction of motion, maximum velocity and maximum acceleration of every single vibration*.

A small movement of the ground is usually composed of a displacement in some definite direction and a rotation about some definite line. Taking any three axes at right angles to one another—say, one vertical and the others horizontal—the displacement may be resolved into three component displacements

* The literature dealing with the theory and construction of seismographs is very extensive. The following books and memoirs may be mentioned as among the more important:

1. Ewing, J. A. Earthquake measurement. *Mem. Sci. Dep.*, Tokyo Univ., No. 9, 1883, pp. 1–92.

2. Galitzin, Prince B. *Vorlesungen über Seismometrie*, 1914, pp. 1–538 (German translation, edited by O. Hecker). Galitzin's original memoirs on seismometry are published in the *Comptes Rendus de la Commission Sismique Permanente* (Petrograd), vols. 1–5, 1902–1912.

3. Knott, C. G. *The Physics of Earthquake Phenomena* (Oxford Univ. Press), 1908, pp. 48–89.

4. Marvin, C. F. A universal seismograph for horizontal motion and notes on the requirements that must be satisfied. *Monthly Weather Rev.* (U.S.A.), Nov. 1907, pp. 1–29.

5. Reid, H. F. Theory of the seismograph. *Californian Earthquake of April* 18, 1906, vol. 2, 1910, pp. 143–190.

6. Walker, G. W. *Modern Seismology* (Longmans), 1913, pp. 1–36.

7. Wiechert, E. Theorie der automatischen Seismographen. *Abhand. der kön. Gesell. Wissen. zu Göttingen, Math. Phys. Kl.*, vol. 2, 1903, pp. 1–128.

Descriptions of numerous seismographs are given by R. Ehlert in *Beitr. zur Geoph.*, vol. 3, 1896–1898, pp. 350–474; H. F. Reid in *Bull. Seis. Soc. Amer.*, vol. 2, 1912, pp. 8–30; and G. W. Walker in *Modern Seismology*, pp. 16–20.

along these lines, and the rotation into three component rotations about the same lines. A complete seismograph should therefore be capable of recording all six components, and no such instrument has yet been devised. It is probable, however, that, except near the epicentre, any movement of rotation is of little consequence, and thus the efforts of seismologists have been concentrated on the construction of instruments that will give a complete account of the displacements in three perpendicular directions, in other words, of instruments that will record the horizontal and vertical movements of the ground during an earthquake. It must be remembered, however, that an instrument designed for registering the horizontal motion in any direction will also be affected by a tilt or rotation of the ground, and that thus there may be some uncertainty whether the recorded movement indicates a displacement only or a displacement complicated by a tilting of the ground.

To describe the numerous seismographs which have been invented would require a volume as large as the present. All that can be attempted in this chapter is to give: (i) an outline of the principles on which the construction of seismographs is based, (ii) an account of a few instruments which are widely used in this country and Japan, and (iii) brief references to other instruments, especially those designed by foreign workers.

The movements of the ground during an earthquake may be of two kinds: (i) in great earthquakes, there may be a large displacement or lurch, either horizontal or vertical, or horizontal and vertical, the range of which may amount to 20 feet and more; and (ii) in all earthquakes, there is a vibratory motion of the ground, the period of the vibrations varying from a fraction of a second in near earthquakes to 20 or 30 seconds in distant earthquakes. The measurement of great lurches is beyond the scope of seismographs*. For the record of the vibratory move-

* It may in some cases be made by a re-triangulation of the district (sect. 84). To measure future movements along the San Andreas fault in California, two series of concrete piers are sunk in the ground along lines at right angles to the fault, each series consisting of four piers, two on each side of the fault-line (Lawson, vol. 1, pp. 152–159). Vertical movements along the Ridgeway fault in Dorsetshire, if any should occur, will be determined by measuring the relative displacement of four brass castings fixed to the rock, two on each side of the fault (H. Darwin, *Rep. Brit. Ass.*, 1900, pp. 119–120).

ments, no single seismograph is sufficient. Different instruments are needed for the registration of near and of distant earthquakes. The minute tremors which precede and accompany a volcanic eruption require a sensitive tremor-recorder or tromometer. In great earthquakes, the ordinary seismographs are usually damaged or thrown out of action, and a fourth form of seismograph, specially adapted for strong and vigorous motion, is necessary.

The Essential Parts of a Seismograph

10. A seismograph consists of at least three parts: (i) the so-called *steady mass*, a certain point or line of which remains, or should remain, steady during the complicated movements of an earthquake; (ii) a frame or *support*, from which the steady mass is suspended, and which partakes in the movement of the ground; and (iii) the recorder, consisting of a lever or beam of light which magnifies the displacement, and a drum or plate (usually in continuous motion) on which the record is inscribed. In modern seismographs, there is also (iv) a *damping device*, the object of which is to check and control the oscillations which the steady mass may acquire during an earthquake.

11. The Steady Mass. Let AB (Fig. 2) represent the vertical axis of a cylinder pivoted about the horizontal line SS, and G the centre of gravity. If the point A be displaced along the line GA, the movement of the cylinder will be one of simple translation in the same direction. If the point A receive a small but sudden displacement in the direction (represented by the arrow) at right angles to the plane SAB, the movement of the cylinder will be one of rotation about a line II, parallel to SS, which meets AB in the point C. The distance AC is such that $AC = k^2/AG$, where k is the radius of gyration of the cylinder about the line SS. The line II is called the *instantaneous axis*, and the point C the *centre of percussion*, with respect to the line SS.

Again, if the cylinder be pivoted about the line $S'S'$ passing through B (Fig. 3), and if the point B receive a small displacement at right angles to the plane $S'BA$, the movement of the cylinder will be one of rotation about the parallel line $I'I'$, which meets BA in the point C' such that $BC' = k^2/BG$.

and thus the magnification of the record is less by unity than the above ratio, or is equal to PS/SI. To obtain a magnifying power of 10, the length PI must therefore be 11 times the distance SI.

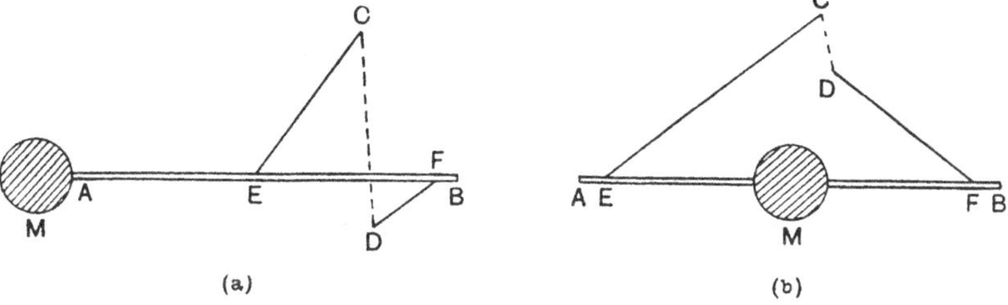

Fig. 5. Support of horizontal pendulum, two wires.

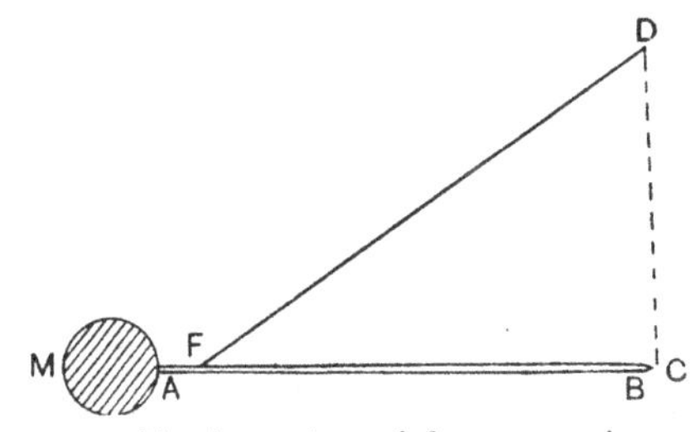

Fig. 6. Support of horizontal pendulum, one wire and one pivot.

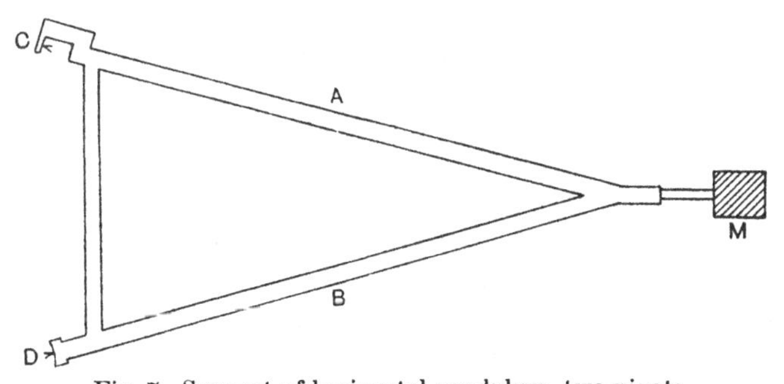

Fig. 7. Support of horizontal pendulum, two pivots.

The magnifying power required varies with the distance and strength of the disturbance. For near earthquakes of moderate strength, one of 2 to 5 is sufficient; for those of destructive intensity, the natural scale is usually employed; for small tremors, such as those which precede or accompany a volcanic eruption, a power of 10 or more is necessary. For recording distant earthquakes, most seismographs have a magnifying power of from 10 to 50, which may with advantage be as high as 100 or even 1000 if the earthquakes should occur near the antipodes.

17. Two methods of registration are in use, (i) mechanical, and (ii) photographic.

In the former, a light lever projects from the pendulum and ends either in a fixed metal point, or, preferably, in a light aluminium pointer hinged about a horizontal axis. This point rests lightly on a sheet of smooth paper or glass covered with a thin coating of soot obtained from a lamp turned on to smoke. When the point is in motion, it removes the soot along a fine white line, the form of which is afterwards fixed by a coating of diluted varnish. The paper or glass may be at rest, if the pendulum have freedom of movement in any direction (sect. 23); but it is usually in motion, the paper being wrapped round a drum driven by clockwork. The paper or glass may be either (i) started by the initial tremors of the earthquake, (ii) in continuous motion, or (iii) in continuous motion, slow at first, but made more rapid on the arrival of the early tremors.

The movement may be recorded photographically in two ways, either (i) directly, as in the Milne seismograph (sect. 26), by a beam of light passing through an aperture in a thin plate prolonging the pendulum, or (ii) by the reflection of a beam of light from a mirror attached to the pendulum, as in the Milne-Shaw seismograph (sect. 27). In either case, the light is concentrated on the photographic paper, which is wrapped round a drum driven by clockwork. When there is no disturbance, the spot of light traces a straight narrow band. The occurrence of an earthquake is indicated by a widening of the band, or, when the paper is driven rapidly, by a series of sinuous waves.

The amount of magnification attained by these methods is limited by the sensitiveness of the photographic paper. A still higher power may be obtained by means of (iii) the galvanometric method employed in the Galitzin seismographs. In this method, several flat induction coils are fixed to the end of the boom of the seismograph and placed between the poles of a pair of strong horse-shoe magnets. When the boom is set in motion, electric currents are generated in the spires of the coils, their intensity being proportional to the angular velocity of the displacement of the boom. These currents are led by two wires to a highly sensitive dead-beat galvanometer, the movements of which are registered by means of a beam of light reflected from a small mirror attached to the moveable coil of the galvanometer. As this method introduces no friction, the magnification attained by it can be raised to 1000 or more. Other advantages of the method are that the recorder can be placed at a considerable distance from the pendulum and at a comparatively short distance from the source of light*.

The principal advantages of mechanical registration are its cheapness, the high velocity that can in consequence be given to the paper with the resulting open and detailed diagrams, and the visibility of the record. Though the effect of the friction between the writing pointer and the paper or glass may be diminished by increasing the weight of the heavy mass, there is always some friction, which limits the power of magnification (to 10 or 20 times). The chief advantages of photographic registration are the entire absence of such friction and the high magnifying power which is attainable. On the other hand, the method is costly, the speed of the paper is usually slow, and, when slow, the diagrams in some instruments show little detail, and the occurrence of the earthquake is not manifested until the record is developed†.

18. Damping Arrangements. The principal defect of an ordinary or free horizontal pendulum is its tendency to take up its own natural period of oscillation, and this defect is especially pronounced when the period of the earthquake-vibrations approaches that of the pendulum. The record of the Derby earthquake of 1903, obtained with an undamped seismograph at Birmingham (Fig. 8), illustrates this defect. It shows the actual earthquake-vibrations superposed on a large undulation clearly due to the swing of the pendulum. In this case, the error is not of great moment,—the amplitude of the vibrations being

* Galitzin, pp. 285–311.

† Marvin (pp. 21–26) suggests that a double system of registration might be employed, one slightly magnified record on smoked paper, the other photographic and magnifying from 100 to 150 times.

merely measured from a curved datum-line. It is more important when, as in the record of a distant earthquake, the waves attain a period of from 12 to 20 seconds. Not only are the waves then unduly magnified, but little trust can be placed on the interpretation of the seismogram.

In modern instruments, this defect is met by some damping device, which opposes the large swings of the pendulum, but which at the same time allows the pendulum to return to its position of equilibrium. Damping of necessity lessens the sensitiveness of a seismograph, and, before applying it, it is necessary to increase the instrument's magnifying power, otherwise minute

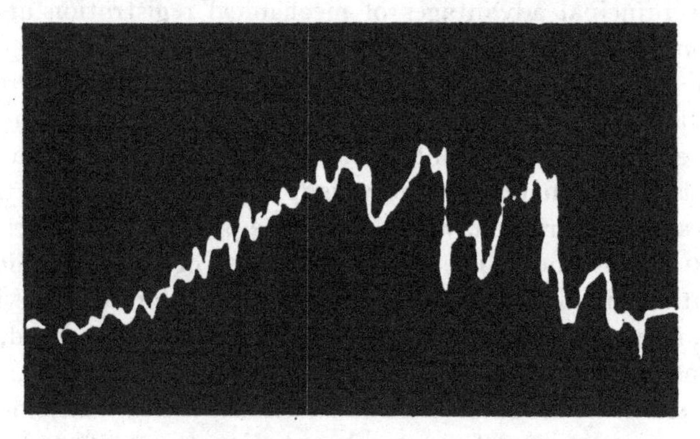

Fig. 8. Record of the Derby earthquake of 1903 obtained with an undamped pendulum.

waves will pass unrecorded. With sufficient magnification, however, instruments may be damped until they become aperiodic, and yet record the slightest tremors as readily as do the free or undamped seismographs. They then record, with a close approach to accuracy, the true nature of the movement during an earthquake.

The principal forms of damping which have been employed are: (i) the friction of the pivots and of the recording apparatus, (ii) the resistance of vanes in a liquid or confined air-space, and (iii) the resistance of electro-magnetic reactions.

Of these forms, the first, which is present in all instruments, is the least desirable. Its effects are difficult to evaluate, and small waves are either quenched altogether or are considerably reduced. In the Darwin bifilar pendulum (sect. 25), the instrument is immersed in oil. In the Wiechert inverted pendulum (sect. 31), an air-damping cylinder is used, by which the instrument can be made as nearly aperiodic as may be desired. The third method is used in the Galitzin and in the Milne-Shaw seismographs (sects. 27, 29). To the steady mass are attached a pair of copper plates, which are free to move between the poles of an electro-magnet. The currents induced in the copper plates offer such a resistance to the motion of the pendulum that it can be made nearly aperiodic*.

Seismographs for Near Earthquakes

19. Ewing Three-Component Seismograph. The pendulum for recording the horizontal motion consists of a solid cylindrical

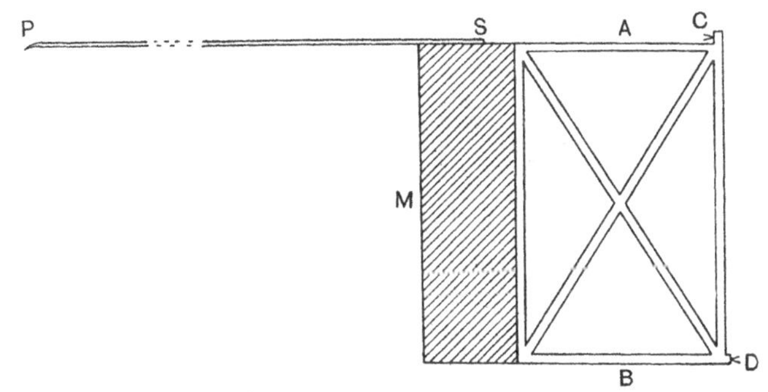

Fig. 9. Ewing horizontal motion seismograph.

brass bob M (Fig. 9) attached to a light rigid brass frame AB. At the lower end D of this frame, there is a conical steel cup; and at the upper end C a projection provided with a vertical V-shaped groove in steel. The cup and groove rest against two sharp points in the frame, adjusted so that the axis of support is inclined slightly to the vertical in the direction of the centre of gravity of the bob. The equilibrium of the pendulum is thus nearly neutral with sufficient tendency to stability to cause the pendulum to return to its original position after displacement, the period of oscillation being about four or five seconds. The multiplying lever consists of a straw SP, terminating in a fine steel point at P. The straw is attached at the end S to a hori-

* Galitzin, pp. 212–216, 265–271.

zontal brass hinge at the top of the bob and its weight is partly supported by a spring wire, so that the point *P* rests lightly on the smoked glass plate. Two exactly similar pendulums are mounted on the same stand at right angles to one another, and the multiplying levers are inclined to the planes of the pendulums so that the writing points lie a short distance apart on the revolving plate. The levers are of such a length that the movements of the ground are multiplied four times in the record.

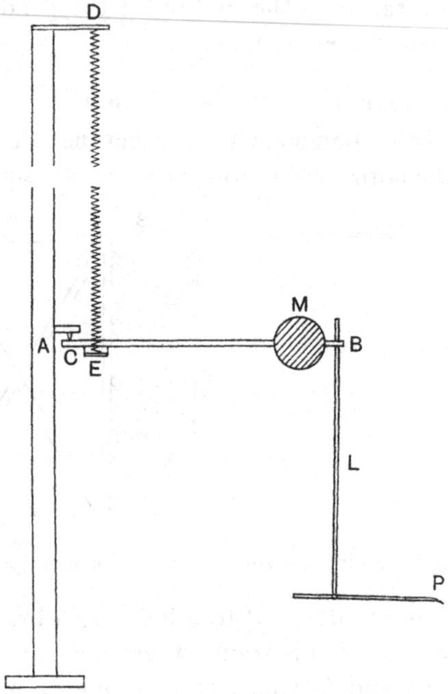

Fig. 10. Ewing vertical motion seismograph.

20. The seismograph designed for recording the vertical component of the motion is a modification of the instrument invented by T. Gray. It consists of a cylindrical brass bob shown in section at *M* (Fig. 10). This is attached to a light rigid rectangular brass frame *AB*, on the upper surface of which at *C* are a conical hole and a V-slot in a line parallel to the axis of the bob. The hole and slot rest against two fine steel points fixed in a stout vertical iron stand. The other support of the pendulum is a pair of spiral springs *DE*, attached at the lower end to a bar *E*,

small hooks *E*, *F* (Fig. 13) on its upper rim, and fastened to two points *C*, *D* in the frame, the axis of support *CD* being inclined at a very small angle to the vertical. The mirror and supporting frame are immersed in oil, and the record is made by a beam of light reflected by the mirror *M* and afterwards concentrated on a sheet of bromide paper wrapped round a revolving drum. Though earthquakes have frequently been registered by the bifilar pendulum, the interesting preliminary tremors are quenched, owing to the immersion of the whole in oil. For measuring slow tilts of the ground, the instrument is

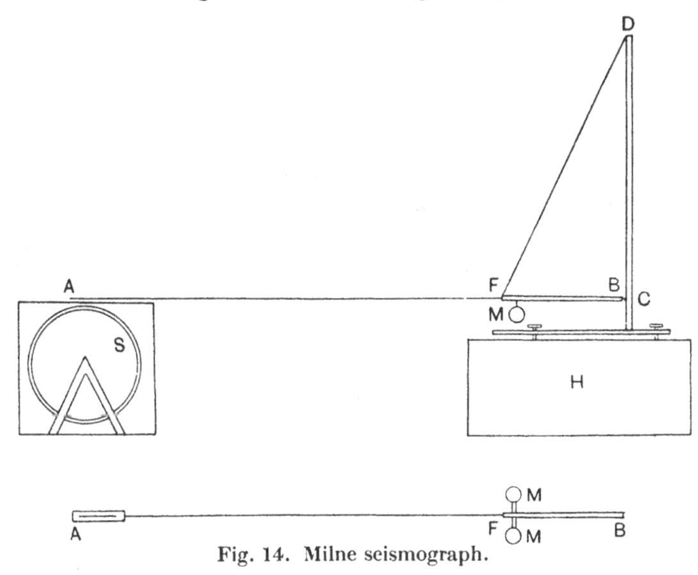

Fig. 14. Milne seismograph.

most effective. It has indeed been found possible to measure a tilt of $\frac{1}{300}$ of a second, an angle which corresponds to raising by one inch the end of a line 1000 miles in length*.

26. Milne Seismograph. The instruments described in this section and the next belong to the second type of horizontal pendulums (sect. 14), the Milne seismograph being one of the smallest and lightest used in the registration of distant earthquakes. The pendulum is supported on a cast-iron column *CD* (Fig. 14) about 20 inches in height. The column is attached to

* *Rep. Brit. Ass.*, 1893, pp. 291–303; 1894, pp. 145–151; *Nature*, vol. 50, 1894, pp. 246–249.

an iron plate which stands on the surface of a column of masonry *H*, the plate being levelled, and if necessary tilted, by three levelling screws passing through it. The beam *AB* consists of a light aluminium rod, about 39 inches long, the end *B* of which terminates in an agate cup that presses against a fine steel point *C* in the column. The other support of the beam is a fine wire ending in a silk thread, *FD*, passing over a small wheel at the top of the column. The bob *M* of the pendulum consists of two small brass balls, each weighing about half a pound, at the ends of a short bar attached to the beam at right angles. The beam ends at *A* in a small aluminium plate (see the section of the beam in the lower part of Fig. 14), in which there is a narrow slit in the direction of the length of the beam. Just beneath the centre of the slit is a perpendicular slit in the top of a case covering the recording drum *S*, round which a strip of bromide paper is wrapped. A ray of light from a lamp is reflected by a mirror down to the drum, which it reaches, after passing through both slits, as a small spot of light. In the early forms of the instrument, the strip of bromide paper was nearly 2 inches wide and about 33 feet long, and was driven at the rate of about $2\frac{1}{3}$ inches per hour, the roll of paper thus lasting for a week. In the later forms, a sheet of bromide paper, $6\frac{1}{4}$ inches wide, is wrapped round a brass cylinder 39 inches in circumference (as shown in Fig. 14). The cylinder revolves once in four hours and at the same time advances one-quarter of an inch along its axis. This gives a more open scale of nearly 10 inches per hour, the sheet of bromide paper lasting about four days. Time-marks are made once an hour by a small shutter which, actuated by an electro-magnet, cuts off the light passing through the fixed slit*.

27. Milne-Shaw Seismograph. The principal defects of the Milne seismograph are its low magnifying power and the absence of damping. A high power is necessary in order that the times of arrival of the two series of preliminary tremors (sect. 147) may be accurately determined, and some kind of damping is desirable so that errors in the identification of the second series may be avoided. Both these defects are removed in the new form devised by J. J. Shaw.

* *Rep. Brit. Ass.*, 1896, pp. 187–188; 1897, pp. 137–145; 1904, pp. 43–44.

The Milne-Shaw seismograph consists of a boom 16 inches long, carrying a mass of one pound. The most important modifications relate to damping and the mode of registration. (i) The boom carries a damping vane which moves in a magnetic field so strong that the pendulum is brought to rest after each oscillation. (ii) The photographic mode of registration is used, but by means of a beam of light reflected by an iridium pivoted mirror, which rotates in an agate setting and is coupled to the free end of the boom. This gives a magnifying power of 300, which is about 40 times that of the Milne seismograph. The beam of light passes first through a vertical cylindrical lens and then, after reflection by the mirror, through a horizontal cylindrical lens, which focusses the light into a small point. Falling on a slit ·003 inch wide in contact with the film, this point gives a definition so fine that waves with a period of only two or three seconds are shown clearly on paper moving at the rate of an inch in three minutes*.

28. Omori Horizontal Pendulum. Omori's pendulum differs chiefly from Milne's in its dimensions and mode of registration. This, being mechanical, involves the use of a heavier steady mass in order to lessen the effect of friction of the recorder. The bob *M* (Fig. 15) in the standard form of the instrument weighs about 30 lb. From the axis of the bob to the socket *C* the distance is nearly 40 inches, that from the point of suspension *D* to the socket *C* about 8 feet. As in other instruments of the same type, the axis of support *CD* is inclined at a very small angle to the vertical in the direction of the steady mass. The recording lever *PQ* is a light aluminium rod, pivoted about an axle *T* connected with the ground. The short arm *Q* ends in a horizontal fork, the prongs of which pass on either side of, and touch, a vertical axle of polished steel. The ends of this axle are fine points which rest in conical sockets, one in the top of the bob at its centre of percussion, the other in a small bridge also attached to the bob. There is thus but little friction between the steel axle and the lever at *Q* when the bob is displaced from its position of equilibrium. At its free end *P*, the lever has a light aluminium pointer, which is hinged about a horizontal axis and rests with light pressure on a sheet

* J. J. Shaw, *Rep. Brit. Ass.*, 1915, pp. 57–58 (also pp. 74–79).

of smoked paper wrapped round the drum *S*. Revolving once an hour, the drum gives to the paper a velocity of 36 inches per hour. At the same time, one end of the axle being screw-cut, the drum is continuously shifted in the direction of its axis, so that the lines traced by the pointer during successive revolutions are separated by about one-sixth of an inch. Time-marks are made once a minute on the smoked paper by an electric time-marker connected with a clock. The period of oscillation of the pendulum can be varied by altering the

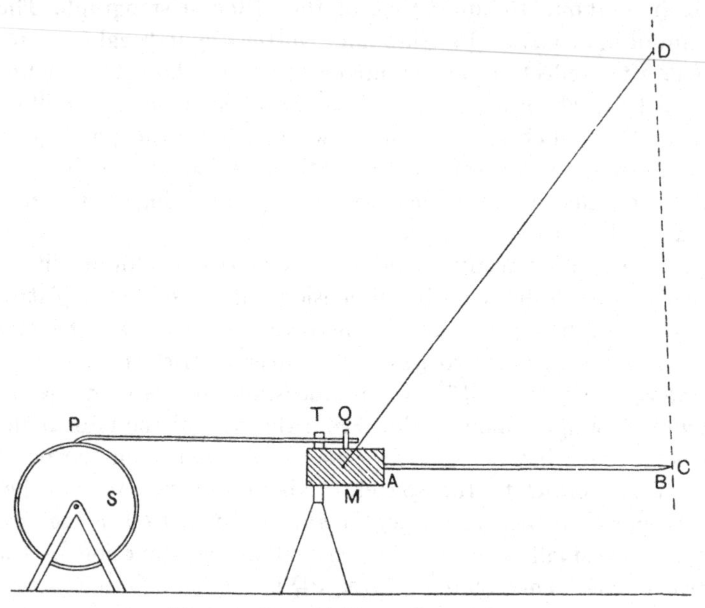

Fig. 15. Omori horizontal pendulum.

position of the upper point of support *D*. It is found convenient, however, not to increase it much above 30 seconds, in order to prevent undue wanderings of the pendulum.

A portable form of the Omori horizontal pendulum differs chiefly in its smaller scale. The weight of the bob is about $6\frac{3}{4}$ lb., the vertical distance *CD* between the points of support and suspension nearly 40 inches, and the horizontal distance between the axis of the bob and the socket *C* nearly 30 inches. The points of support and suspension are attached to a strong cast-iron stand bolted to a column of stone. As a rule, the period of oscillation is kept at about 20 seconds, though, if desired, it can be raised to considerably above this value without affecting the stability of the pendulum*.

29. Galitzin's horizontal-motion seismograph is suspended, as in Hengeller's and Zöllner's pendulums (Fig. 5, *a*), by two steel wires, the mass weighing about 15½ lb. The boom is prolonged beyond the mass, and near the end are fixed two pairs of powerful horse-shoe magnets, one of each pair above, and the other below, the boom. A copper plate fixed to the boom passes between one pair of magnets, the distance between being adjustable so as to render the pendulum aperiodic. At the end of the boom are fixed flat induction coils, which pass between the other pair of magnets, and are connected with the registering galvanometer (sect. 17).

Galitzin's vertical-motion seismograph is based on the principle suggested by Gray (sect. 20), the rod carrying the heavy mass being replaced by a triangular frame in order to avoid distortion by bending. The damping and registering arrangements are similar to those employed in the horizontal-motion seismograph.

30. Common Pendulums. One of the most widely used forms of the common pendulum is the Vicentini microseismograph. In the instrument erected in the University of Siena, the bob of the pendulum weighs about 1 cwt. and is supported by three chains, united at their upper ends in a brass cap. An iron wire is attached at one end to this cap, and at the other to a screw in a strong iron bracket driven into the wall of the observatory. The pendulum is nearly 5 feet long, and its period of oscillation, when connected with the recording levers, is about 2·4 seconds. The bob is surrounded by an iron ring carrying three screws, arranged so as to prevent large oscillations of the pendulum that might injure the delicate recording apparatus. By a light vertical lever projecting downwards from the bob, the movements of the pendulum are magnified 16 times. Those of the lower end of the vertical lever are further magnified five times by a pair of light horizontal levers, which give the components

* F. Omori, *Journ. Coll. Sci.*, Imp. Univ. Tokyo, vol. 11, 1899, pp. 124–130.

of the motion in directions at right angles to one another. The levers are made of thin aluminium plate, one being bent at right angles so that the fine glass fibres, in which the long arms terminate, trace the components of the horizontal motion on the same sheet of smoked paper driven at the rate of nearly 5 inches per hour.

Many of the pendulums used in Italy are much longer than the preceding and are provided with heavier masses. For instance, the Agamennone microseismometrograph, installed in the tower of the Collegio Romano, Rome, is 54 feet long and carries a bob weighing 4 cwt. A Cancani microseismometrograph erected at the observatory of Catania has a mass weighing 6 cwt. and is 85 feet long*.

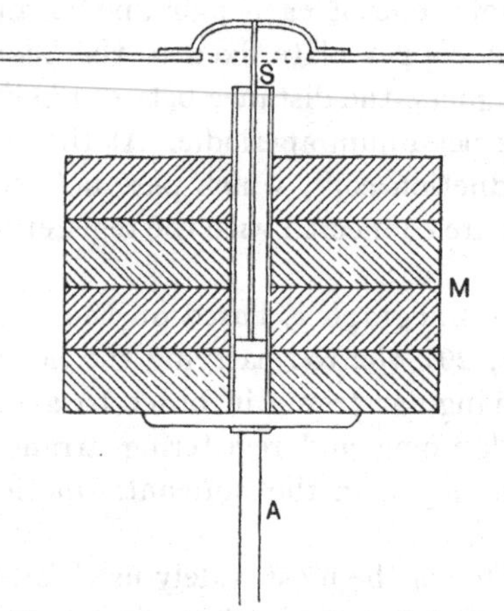

Fig. 16. Marvin inverted pendulum.

Pendulums of such great length are obviously difficult to instal, and the rigidity of their supports is uncertain. Moreover, in order to save the recording apparatus from injury, it is necessary to place stops round the steady mass, and, when these are struck, the subsequent record becomes valueless.

31. Inverted Pendulums. Though used as early as 1841 by J. D. Forbes, it is only lately that the inverted pendulum has been widely employed as a seismograph.

In the Marvin inverted pendulum, a steady mass *M* (Fig. 16) of more than 2000 lbs. (or nearly a ton) is carried by a strut *A*, made of ordinary iron pipe and terminating at the lower end in a universal pivot of ribbon-steel. The pendulum is rendered astatic by the reaction of a cylindrical steel spring rod *S*, fixed at its upper end, and provided at the lower end with a small cylinder which justs fits into a tube bored centrally through the

* Fuller accounts of the instruments designed by Vicentini, Agamennone and Cancani are given in *Rep. Brit. Ass.*, 1896, pp. 40–47.

steady mass and which touches it at the centre of percussion of the mass.

In the Wiechert astatic inverted pendulum, the steady mass consists of iron plates weighing altogether about 1 ton. This is carried by a strong iron rod, which, at its lower end, is supported on two pairs of Cardan springs (or steel bands) at right angles to one another. The springs enable the rod to turn with hardly perceptible friction about any horizontal axis, and their action causes the pendulum after disturbance to return to its position of equilibrium. Two light arms and a system of multiplying levers connect the steady mass with the writing pointers, which are fine glass rods with rounded tips resting on a sheet of smoked paper. The damping apparatus is referred to in sect. 18. With this instrument, it is possible to magnify the movements of the ground about 500 times. A seismograph has also been constructed with a mass of nearly 17 tons and a magnifying power of 2200*.

* J. D. Forbes, *Trans. Roy. Soc. Edin.*, vol. 15, 1844, pp. 219–228; Marvin, pp. 13–15; E. Wiechert, *Beitr. zur Geoph.*, vol. 6, 1904, pp. 435–450.

CHAPTER III

NATURE AND INTENSITY OF EARTHQUAKE-MOTION

Nature of Earthquake-Motion

32. In a great earthquake, there co-exist many different types of vibrations, of which personal impressions alone or seismographs alone would give an imperfect representation. Some movements elude observation by reason of their small amplitude, others owing to the length of their period. There are, again, vibrations so minute and so rapid that they may escape instrumental record and yet be perceptible to the ear as sound. Lastly, in the central district of a great earthquake, the surface of the ground may be crossed by waves that are so large and travel so slowly that they are clearly visible as they move.

33. Nature of Earthquake-Motion: Personal Impressions. The first sign of an earthquake is usually a low rumbling sound, so low that to some observers, not otherwise deaf, it is quite inaudible. The sound grows gradually louder, and with it there become perceptible faint but rapid tremors, like those experienced during the passage of a heavy train or waggon, about four or five tremors occurring to the second. Both sound and tremors increase in strength until, after a few seconds, they merge more or less rapidly into vibrations of considerable range and duration, not more than two or three per second, while, in strong earthquakes, each vibration may last a second or even more. With these vibrations, deep explosive crashes are often heard in the midst of the rumbling sound. There may be variations of strength in the vibrations, but, except in great earthquakes, when once the maximum is reached, the vibrations and sound usually decrease in strength, the shock sometimes ending with a tremulous motion, while the sound may last for a second or two after the vibrations have become insensible.

In the simplest case, the three phases of a moderately strong earthquake may be represented by the curve in Fig. 17, *a*, the part *AB* denoting the preliminary tremors, *BC* the principal portion, and *CD* the concluding tremors.

In very slight earthquakes, the principal portion may be absent, and a weak tremor may alone be felt. In some slight earthquakes, the preliminary or concluding tremors may be omitted, usually the former, in which case the sensible shock begins with a single prominent vibration, like the thud of a falling body, followed by brief tremors such as a fall would produce in a building.

On the other hand, in a great earthquake, such as that of

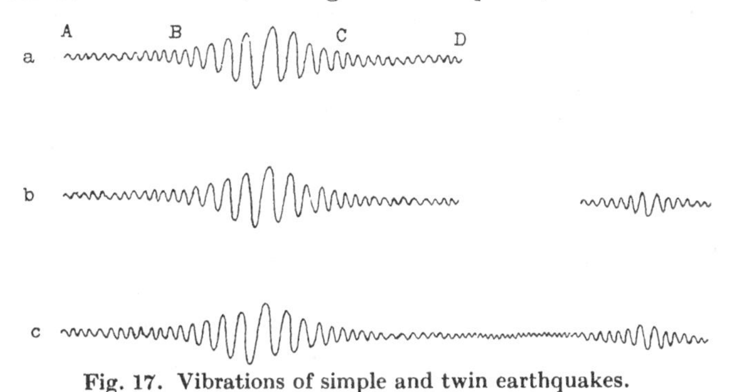

Fig. 17. Vibrations of simple and twin earthquakes.

Kangra in 1905, the vibratory motion within the inner isoseismals may be complicated by sudden lurches as if the ground moved as a whole*.

34. Earthquakes have been divided into undulatory, sussultatory or vorticose shocks according as the movement is nearly horizontal, nearly vertical, or frequently changing in direction or rotatory. Such vibrations, however, have no physical meaning; they depend only on the position of the observer with reference to the epicentre. The shock is vorticose when the observer is near a large focus, the earth-waves reaching him in succession from different parts of the focus. It is sussultatory at places near the epicentre and undulatory at some distance from it.

A more natural classification, and one depending on the nature of the originating movement, is that into simple, twin and complex earthquakes. (i) In *simple* earthquakes, the vibrations

* C. S. Middlemiss, *Mem. Geol. Surv. India*, vol. 38, 1910, pp. 306–317.

increase in intensity to a maximum, and then as a rule die away (Fig. 17, *a*), the duration of the shock varying from a second or two to many seconds. (ii) In *twin* earthquakes, the movement is similar to that in simple earthquakes, but it is repeated after the lapse of two or three seconds (Fig. 17, *b*). Sometimes, weak tremors are felt during the interval between the main parts of the shock (Fig. 17, *c*); but, at some distance from the epicentre, these tremors become imperceptible and the shock consists of two detached parts. In moderately strong twin earthquakes, such as those of Great Britain, the duration ranges from seven or eight to about 15 seconds; in strong twin earthquakes, such as the Charleston earthquake of 1886 or the Messina earthquake of 1908, the duration may be a minute or even more. (iii) Among *complex* earthquakes are the greatest of all shocks. They last much longer than those of the other classes, some, like the Californian earthquake of 1906, for three or four minutes, during which there are many variations of strength and rapid changes of direction.

35. Nature of Earthquake-Motion as revealed by Seismographs. The present section will be confined to the nature of the earthquake-motion at places within the disturbed area. In Chapter IX, the nature of the motion registered by seismographs at very great distances will be considered*.

In earthquake-motion as registered by seismographs, there are again three phases: (i) the *preliminary tremor*, in which the vibrations are of very small amplitude and generally short period; (ii) the *principal portion*, in which the vibrations are of considerable amplitude and long in period; and (iii) the *end-portion*, consisting of vibrations which may escape observation owing to their diminished amplitude and comparatively long period.

It is sometimes convenient to use the term *ripples* to denote minute vibrations, the period of which is a small fraction of a

* So far as precision has been attained, our knowledge of the nature of earthquake-motion is almost entirely due in the first place to the work of British seismologists in Japan, and especially to that of J. Milne, J. A. Ewing and T. Gray. Their results are to be found in the *Transactions of the Seismological Society of Japan* (vols. 1–16, 1880–1892). If the references in the present chapter are mainly to the memoirs of later seismologists, the reason is that, with the lapse of time, more observations have become available.

second, and the term *slow undulations* to denote comparatively gentle movements with periods of nearly, or even more than, a second.

The limits of the principal portion are usually well defined at places within a moderate distance from the epicentre, but at great distances they become indefinite. In this phase, there is as a rule no single large vibration which stands out prominently from the rest, though an example of such a vibration is illustrated in Fig. 19. In most earthquakes, there are several or many movements of nearly equal amplitude; in some, there may be several maxima with intervals of comparative rest between.

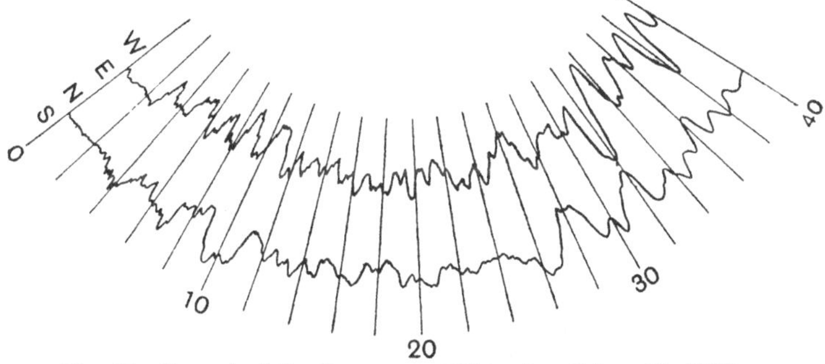

Fig. 18. Record of the Japanese earthquake of Jan. 15, 1887.

In most of the earthquakes registered at any station, the vertical component is absent or imperceptible, an omission which is probably due to the distance of the seismic focus. Of the earthquakes registered at Tokyo from Sep. 1885 to Sep. 1887, Sekiya finds that only 28 per cent. were accompanied by vertical motion; and the vertical vibrations, when they are recorded, are smaller in amplitude and last as a rule for only part of the time during which the horizontal components are sensible*.

36. Examples of Earthquake-Records. The four examples of earthquake-records here given were all obtained in Japan. Owing to the large size of the diagrams and the long duration of the first two, it is only possible in these cases to reproduce the more characteristic parts of the records.

The first example given is that of the destructive earthquake

* *Trans. Seis. Soc. Japan*, vol. 12, 1888, pp. 83–106.

of Jan. 15, 1887, the epicentre of which was about 35 miles south-west of Tokyo. In Fig. 18 are reproduced the two horizontal components during the first 40 seconds of the motion, as given by an Ewing seismograph at Tokyo. The radial lines mark alternate seconds of time. The earthquake began as usual with short-period tremors. During the third second, the principal portion followed abruptly with a vigorous horizontal motion in the N.W.–S.E. direction, or at right angles to the direction of the epicentre from Tokyo. This was accompanied by a considerable vertical displacement. Both horizontal and vertical motions then continued with great activity, the greatest

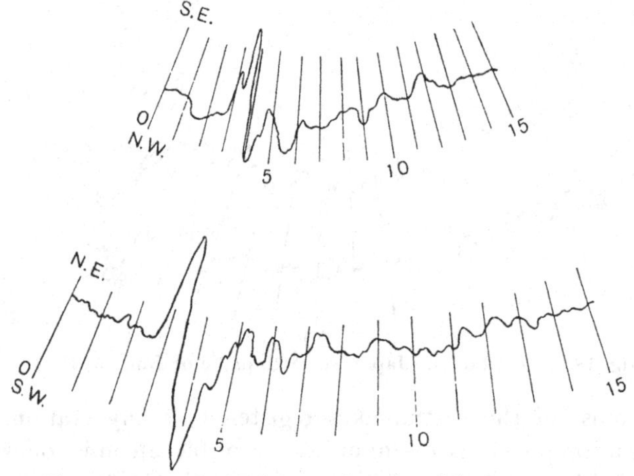

Fig. 19. Record of Tokyo earthquake of June 20, 1894.

horizontal movement of 7·3 mm. occurring from the 33rd to the 34th second, with a complete period of 2 seconds. The vertical motion died out practically after 71 seconds, but horizontal displacements, occasionally of considerable magnitude, continued for more than a minute longer. Two points in particular should be noticed about this diagram: (i) the short-tremors characteristic of the first phase of the movement did not cease with that phase, for they were superposed on many of the early undulations of the principal portion. (ii) The undulations of the two components as shown in Fig. 18, were sometimes in the same direction, but often in opposite directions, showing that the resultant movement must have taken place in many different directions.

The second example (Fig. 19) represents the horizontal components of the semi-destructive Tokyo earthquake of June 20, 1894. This was obtained with an instrument specially designed for registering strong earthquakes and erected in the University Observatory at Tokyo. The diagram illustrates the nature of a strong earthquake originating at a short distance from the earthquake-station. The preliminary tremors in this case lasted about 3 seconds*. The motion then suddenly became violent, the ground moving 37 mm. in the direction N. 70° E., followed by a counter-movement of 73 mm. (the greatest experienced during the earthquake), and this again by a return movement of 42 mm. The period of this prominent undulation was 1·8 seconds, the maximum acceleration being therefore 444 mm. per second per second (sect. 2, footnote). After this, the vibrations were all comparatively small, dying away, though with some variations in strength, in 4½ minutes from the beginning of the earthquake. The vertical motion began at the same time as that in the horizontal direction, the maximum vertical displacement of 10 mm. occurring concurrently with the great horizontal undulation. The vertical motion as usual ceased much more rapidly than the horizontal, being imperceptible after about half a minute. In this earthquake, the principal portion consisted of little more than the one prominent vibration, the end-portion beginning before the lapse of the 15 seconds represented in Fig. 19.

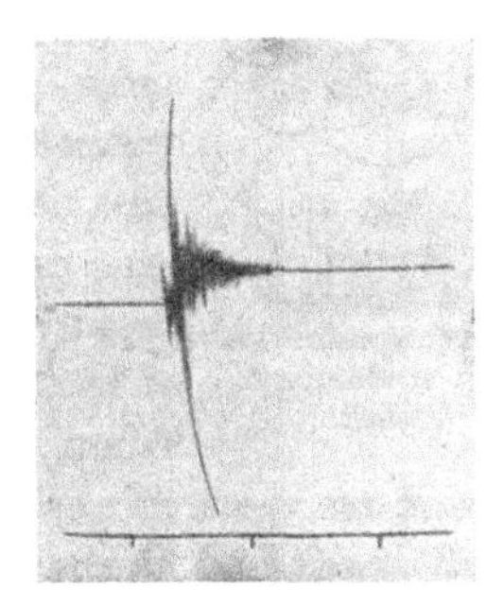

Fig. 20.
Record of Japanese earthquake of July 27, 1905.

Fig. 20 shows the characteristic features of a sharp local shock registered by a horizontal tremor-recorder at Mount Tsukuba on July 27, 1905. These features are the shortness of the preliminary tremor and the occurrence of the maximum vibration of both preliminary tremor and principal portion at the be-

* According to the diagrams of ordinary seismographs, the preliminary tremors lasted more than 10 seconds.

ginning of the respective phases. In this case, the preliminary tremor lasted 7·9 seconds, and the maximum motion of this phase was ·04 mm. and of the principal portion ·34 mm. Fig. 21 is the record of a Japanese earthquake given on a stationary plate by an Ewing duplex-pendulum seismometer*.

37. Visible Undulations. Besides the earth-waves which are felt and heard and which travel with great velocity, there are others present in great earthquakes which cross the central areas and are readily visible owing to their magnitude and comparatively small velocity. In some ways, they seem to resemble ordinary water-waves. The surface of the ground as they pass is said to be like that of a storm-tossed sea, except that the undulations move more rapidly. Sometimes, as in the Assam earthquake of 1897, the waves come from opposite directions. When they meet, the earth rises and water and sand are ejected from the soil; as the waves pass by, the ground falls back and opens out in long fissures. It is difficult to form exact estimates of the magnitude of such waves, but the evidence collected after several earthquakes agrees in placing their height at a foot or even more and their length at about 30 feet. They travel quickly (faster than a man can walk but not so fast as he can run, according to one observer in Assam), but the mere fact of their visibility shows that their velocity must be much less than that of ordinary earth-waves†.

Fig. 21. Record of a Japanese earthquake on a stationary plate.

ELEMENTS OF EARTHQUAKE-MOTION

38. The elements of earthquake-motion referred to in sects. 38–41 are all determined from seismographic records. Similar elements, calculated in part from the overthrow of columns and relating to great earthquakes only, will be described in sects. 43, 44. In both cases, the observing stations are supposed to be at no great distance from the epicentre.

* S. Sekiya, *Trans. Seis. Soc. Japan*, vol. 11, 1887, pp. 79–88, 175–177; S. Sekiya and F. Omori, *Journ. Coll. Sci.*, Imp. Univ. Tokyo, vol. 7, 1894, pp. 1–4, and *Publ. Eq. Inv. Com.*, No. 4, 1900, pp. 35–38; F. Omori, *Publ. Eq. Inv. Com.*, No. 22 A, 1908, pp. 25–26.

† Dutton, pp. 264–268; Oldham, pp. 5, 7, 20, 26, 27 and 37.

39. Period of the Vibrations. As a rule, an earthquake begins with tremors of which there may be five or more to the second. In the earthquakes recorded at Miyako from 1896 to 1898, the mean period of the ripples in the first phase of the motion was ·08 second, or at the rate of 12½ vibrations per second. The most rapid tremors ever recorded are probably those of one of the after-shocks of the Mino-Owari earthquake of 1891 at Midori, the period of which was only ·023 second. As this corresponds to 43½ vibrations per second, it is clear that tremors so rapid would be sensible to most persons as sound.

In the principal portion, the period of the vibrations usually ranges from half a second to a second, but is sometimes, as in the Tokyo earthquake of Jan. 15, 1887, as great as 2½ seconds. In the earthquakes recorded at the Hitotsubashi observatory (Tokyo) in the years 1887–1889, the average period in the horizontal component was ·76 second, and in the vertical component ·53 second. In those recorded at the Hongo observatory (Tokyo) during the same years, the period in the horizontal component was usually about ·57 second, but other periods of ·22 second, 1·18 seconds and 2·20 seconds also occurred; while that in the vertical component was usually ·20 second, with others of less frequent occurrence of ·41 second and ·63 second.

In the earthquakes recorded at Miyako from 1896 to 1898, slow undulations are especially prominent in the principal and end portions, though they occur also in the first phase. Their average period is nearly the same in all three phases, that of the E.–W. component being respectively 1·1, 1·3 and 1·3 seconds, and that of the N.–S. component 1·0, 1·0 and ·94 seconds. The average period of the ripples was nearly the same in all three components, being ·08 second in the first and third phases, and slightly greater, namely ·10 second in the principal portion*.

40. Range of the Vibrations. The range (or double amplitude) of the vibrations is usually less than 1 mm. In the semi-destructive Tokyo earthquake of June 20, 1894 (Fig. 19), the range of the largest vibration was 73 mm. In great earthquakes, even

* S. Sekiya, *Trans. Seis. Soc. Japan*, vol. 12, 1888, pp. 83–106; F. Omori, *Journ. Coll. Sci.*, Imp. Univ. Tokyo, vol. 11, 1899, p. 147, also *Boll. Soc. Sis. Ital.*, vol. 2, 1896, p. 181, and *Publ. Eq. Inv. Com.*, No. 11, 1902, pp. 51–55, 63–64; F. Omori and K. Hirata, *Journ. Coll. Sci.*, Imp. Univ. Tokyo, vol. 11, 1899, pp. 191–193.

this amount is exceeded; but, in such cases, the overthrow of the seismographs prevents the registration of the exact amount. At Nagoya, the maximum range of the Mino-Owari earthquake of 1891 must have been about 223 mm. or 9 inches*.

In the earthquakes recorded at the Hitotsubashi observatory (Tokyo) in 1887–1889, the average range of the largest vibration was ·70 mm. in the horizontal, and ·22 mm. in the vertical, component. In those recorded at the Hongo observatory (Tokyo) during the same years, the corresponding figures were ·79 mm. and ·22 mm. Of the 433 earthquakes recorded during the years 1885–1897 at the Central Meteorological observatory of Tokyo, the range was measured in 366 cases. In all but seven of these, the range of the largest vibration was less than 6 mm., and in about 60 per cent. of the total number, it did not exceed half a millimetre. In four earthquakes, the maximum range was considerable, namely, 22·8, 28·4, 41·0 and 76·0 mm.†

41. Maximum Acceleration of the Vibrations. In slight earthquakes, the maximum acceleration is usually not more than 5 or 10 mm. per sec. per sec., the average for 64 earthquakes recorded at Hitotsubashi (Tokyo) being 20 mm. per sec. per sec. A value as high as 50 mm. per sec. per sec. is somewhat rare; and it is only in strong earthquakes that it exceeds 200 or 300 mm. per sec. per sec. In the semi-destructive Tokyo earthquake of June 20, 1894, the maximum acceleration was 444 mm. per sec. per sec. at the Hongo observatory, and 900 mm. per sec. per sec. at that of Hitotsubashi‡. The corresponding figures for certain destructive earthquakes are given in sect. 53.

Intensity of Earthquake-Motion

42. Lower Limit of Sensible Motion. The least value of the maximum acceleration that is sensible to the unaided senses may be determined in two ways: (i) by the examination of the smallest values of the maximum acceleration of sensible earth-

* The period of the largest vibrations at Nagoya seems to have been about 1·3 seconds, while the maximum acceleration, as determined by the overthrow of columns (sect. 43) was about 2600 mm. per sec. per sec. The formula $f = 4\pi^2 a/T^2$ (sect. 2, footnote) gives the range of $2a$ equal to 223 mm.; see F. Omori, *Publ. Eq. Inv. Com.*, No. 4, 1900, p. 17.

† F. Omori, *Publ. Eq. Inv. Com.*, No. 11, 1902, pp. 51–55, 94–95.

‡ F. Omori, *Boll. Soc. Sis. Ital.*, vol. 2, 1896, pp. 189–191.

quakes, deduced from seismographic records, and (ii) from the values determined at places which are known to be on or near the boundary of the disturbed area of an earthquake.

(i) In his examination of the records of many sensible local earthquakes recorded at Mount Tsukuba, Omori found that the range (or double amplitude) in the case of earthquakes unaccompanied by sound was seldom less than ·013 mm., though it was equal to ·01 mm. in 14 earthquakes accompanied by sound. Taking the latter value as the lower limit of the range, and about one-tenth of a second as the period, the corresponding value of the maximum acceleration would be 17 mm. per sec. per sec.

(ii) Omori has also examined the records of earthquakes registered in Tokyo in which that city lay on or close to the boundary of the disturbed area. In 23 earthquakes recorded at the observatory of Hitotsubashi, the mean values of the range, period and maximum acceleration were respectively ·47 mm., ·74 second and 17·0 mm. per sec. per sec. For 22 earthquakes recorded at the observatory of Hongo, the corresponding values were ·35 mm., ·64 second and 16·9 mm. per sec. per sec.

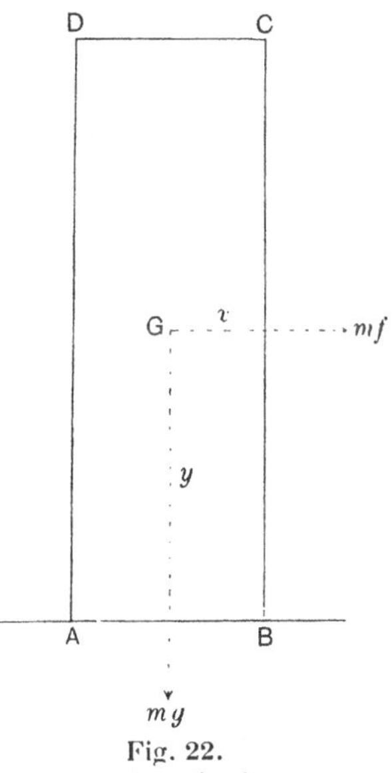

Fig. 22.
Overturning of columns.

Thus Omori concludes that a maximum acceleration of 17 mm. per sec. per sec. is the lower limit of motion that is sensible without instrumental aid*.

43. Determination of the Maximum Acceleration by means of the Overthrow of Columns. In great earthquakes, such as that of Mino-Owari (Japan) in 1891, the range and maximum acceleration in the meizoseismal area are sufficient to put any ordinary seismograph out of action. In these cases, the overthrow of columns of known dimensions provides under certain

* *Publ. Eq. Inv. Com.*, No. 11, 1902, p. 60; No. 22 A, 1908, pp. 37–39.

conditions a satisfactory substitute. Thus, let *ABCD* (Fig. 22) represent a cylindrical column resting on the ground at *AB*, y the height of the centre of gravity G above the ground, and x its horizontal distance from the edge B. If the period of the earthquake-motion be not very small compared with the period of rocking of the column, the column will move with the ground until the acceleration applied to it is sufficient to overthrow it. If, then, f be the maximum acceleration in the direction of the arrow, the least value of f that will overturn the column is given by the equation

$$mfy = mgx,$$

or

$$f = gx/y,$$

the column being overthrown in the same direction as that in which the movement of the ground at the time is taking place. The last equation is known as West's formula.

Experiments to test the accuracy of West's formula have been made by Milne and Omori. Columns were placed on a shaking table, the movements of which were caused to simulate those of the ground during an earthquake by conforming to the law of simple harmonic motion. The experiments show that the formula gives very nearly accurate results if (i) the period of the earthquake-motion be comparable with that of the columns when rocking, and if (ii) the amplitude of the earth's motion be not small. Thus, West's formula may be trusted to give accurate lower limits to the maximum acceleration experienced during a great earthquake*.

An interesting confirmation of this result is furnished by the use of columns of different materials, for West's formula depends on the dimensions only of the overturned body and not on its mass. Columns of brick and iron and wooden boxes, with the same external dimensions, furnished nearly the same values of the maximum acceleration†.

* If f be the value of the maximum acceleration required to overthrow a column as given by West's formula and F the value determined by experiment, Omori finds that the mean value of the ratio $f:F$ is 1·07 : 1 (*Publ. Eq. Inv. Com.*, No. 4, 1900, p. 136).

† J. Milne, *Trans. Seis. Soc. Japan*, vol. 8, 1885, pp. 1–82; J. Milne and F. Omori, *Seis. Journ.*, vol. 1, 1893, pp. 59–86; F. Omori, *Seis. Journ.*, vol. 2, 1893, pp. 119–122; *Boll. Soc. Sis. Ital.*, vol. 2, 1896, pp. 189–200; and *Publ. Eq. Inv. Com.*, No. 4, 1900, pp. 69–141, and No. 12, 1902, pp. 8–27; *Bull. Eq. Inv. Com.*, vol. 4, no. 1, 1910, pp. 1–31.

44. Maximum Acceleration in Destructive Earthquakes. Omori has estimated the maximum horizontal acceleration in several destructive earthquakes by observations on overturned bodies. In the Mino-Owari earthquake of 1891, he found the maximum acceleration to be 2500 mm. per sec. per sec. at Fukui, 2600 at Nagoya, 3000 at Gifu and Ogaki, 4000 at Kasamatsu, and more than 4300 at Iwakura and Komaki. In the Californian earthquake of 1906, the range of motion was about 4 inches and the period of vibration about 1 second, the corresponding acceleration being about 2000 mm. per sec. per sec. Omori estimated the maximum acceleration at Messina during the earthquake of 1908 at approximately the same figure.

With these figures may be compared those which Oldham has given for the maximum horizontal acceleration during the Assam earthquake of 1897. He estimates it as 3000 mm. per sec. per sec. at Cherrapunji, 3600 at Gauhati, Shillong and Sylhet, and 4200 at Goalpara. At Shillong, Gauhati and other places in the epicentral tract, the actual value must have been far higher. At these places, stones were projected upwards, showing that the vertical component of the maximum acceleration must have been greater than that of gravity, which is 9600 mm. per sec. per sec.*

45. Relations between the Nature of the Ground and the Intensity of the Shock. In all earthquakes, the shock is felt more severely on soft ground than on hard compact rock. Milne, in his seismic survey of Tokyo made in the years 1884–1885, showed that the period of the more prominent earthquake-vibrations was greater on soft than on comparatively hard ground, that the range of motion was greater in moderately strong earthquakes though not always in slight earthquakes, and that the maximum acceleration was also greater.

Omori has also compared the records of two observatories in Tokyo. The observatory of Hongo is situated in the higher part of the city where the ground is of hard clay, that of Hitotsubashi

* F. Omori, *Boll. Soc. Sis. Ital.*, vol. 2, 1896, pp. 192–197; *Publ. Eq. Inv. Com.*, No. 4, 1900, pp. 13–16; *Bull. Eq. Inv. Com.*, vol. 1, 1907, p. 19, and vol. 3, 1909, p. 40; Oldham, pp. 78–79, 129–134. The maximum acceleration during lateral vibrations of railway-carriages was found by Omori to be sometimes as much as 2000 mm. per sec. per sec. (*Bull. Eq. Inv. Com.*, vol. 4, no. 3, 1912, p. 97).

stands on low ground consisting of very soft soil. Taking, first, the non-destructive earthquakes from Sep. 1887 to June 1889, the average period of the principal vibrations is ·63 second at Hongo and ·87 second at Hitotsubashi, the average maximum amplitude ·56 and 1·07 mm., and the average maximum acceleration 20·4 and 24·9 mm. per sec. per sec. In the semi-destructive Tokyo earthquake of June 20, 1894, the period of the principal vibration was 1·3 seconds at Hongo and 1·7 seconds at Hitotsubashi, the maximum horizontal range 73·0 and 130·0 mm., and the maximum acceleration 444 and 900 mm. per sec. per sec.

46. The most detailed study of the relation between the nature of the ground and the destructive power of the shock is that made by H. O. Wood of the damage wrought at San Francisco by the Californian earthquake of 1906. The city of San Francisco lies between about 1 and 9½ miles east of the great fault, the movement along which gave rise to the earthquake (sect. 84). On the whole, the intensity of the shock decreased with increasing distance from the fault, but it was subject to many variations evidently connected with the nature of the ground.

The shock was slightest, resulting in the occasional fall of chimneys, in a few small areas, which are invariably those occupied by hard rock (sandstone, chert, etc.) with a level surface. A higher degree of intensity, corresponding to general, but not universal, fall of chimneys, with cracks in masonry and brickwork, marks ground consisting of hard rock with an inclined surface or hard rock with a thin coating of soil. On thick beds of naturally formed alluvium, old and well-compacted, brickwork and masonry were badly cracked, some gables were thrown down, and chimneys generally were destroyed. The worst damage occurred on newly made land, especially on that filling up a marsh or creek. Here, brick and frame buildings generally collapsed, the surface of the ground was thrown into broad undulations, sewers and water-mains were broken*.

* J. Milne, *Trans. Seis. Soc. Japan*, vol. 10, 1887, pp. 1–36; vol. 13, 1890, pp. 41–89; C. Davison, *The Hereford Earthquake of Dec.* 17, 1896 (1899), pp. 276–278; F. Omori, *Publ. Eq. Inv. Com.*, No. 11, 1902, pp. 57–58, and *Boll. Soc. Sis. Ital.*, vol. 2, 1896, p. 191; H. O. Wood, *The Californian Earthquake of April* 18, 1906 (Lawson), vol. 1, 1908, pp. 220–245. M. Baratta gives roughly the same sequence as H. O. Wood for the Messina earthquake of 1908 in *Catastrophe Sismica Calabro-Messinese*, 28 *dicembre* 1908, 1910, p. 230.

47. Comparison between the Motion on the Surface and in Pits. To the unaided senses, there is a perceptible difference between the intensities of an earthquake at the surface and in mines, a shock which is strong on the surface being hardly felt or not felt at all at some depth below. In the Hereford earthquake of 1896 and the Derby earthquake of 1903, the distance from the epicentre of the farthest mine in which the shock was felt was only one-third of the mean radius of the isoseismal 4 (sect. 51). In the meizoseismal area of the Riviera earthquake of 1887, the shock was very weak or not felt at all in the tunnels of the Nice to Genoa railway, and none of the tunnels was damaged in the slightest degree.

There is also a marked difference in the strength of a shock in slight hollows or excavations at the surface and on the adjoining ground. For instance, in the central tract of the Mino-Owari earthquake of 1891, the railway-lines were everywhere more or less disturbed except in small cuttings. Even if the cuttings were not more than 20 or 50 feet in depth, the rails and sleepers were unmoved. At Dharmsala, during the Kangra earthquake of 1905, one house surrounded on several sides by higher spurs and ridges, was spared while the destruction of buildings on elevated ground was either total or nearly so*.

48. The reason for this comparative immunity from the shock in hollows and mines is furnished by some interesting experiments of Sekiya and Omori made at Tokyo in the years 1887 to 1889. The records of two similar seismographs, one at the bottom of a pit 18 feet deep, the other on the surface within a few yards from the pit, were compared for thirty earthquakes, of which three were severe and the rest slight. For the latter, it was found that the average amplitude, maximum velocity and maximum acceleration differed but little on the surface and in the pit, each being slightly greater on the surface. In the large undulations of the severe earthquakes, the three elements again differed but slightly, the greater magnitude being in each case on the surface. It was in the ripples of these earthquakes that

* C. Davison, *Hereford Earthquake*, pp. 278–280, and *Quart. Journ. Geol. Soc.*, vol. 60, 1904, pp. 227–228; A. Issel, *Boll. del R. Com. Geol. d' Italia*, *anno* 1887, pp. 117–120; J. Milne, *Seis. Journ.*, vol. 1, 1893, p. 133; C. S. Middlemiss, *Mem. Geol. Surv. India*, vol. 38, 1910, p. 21.

the greatest difference was manifested, the amplitude being about twice, the maximum velocity three times, and the maximum acceleration five times, as great at the surface as in the pit. Owing to their much shorter period, the ripples at the surface have a maximum acceleration from five to ten times as great as that of the large undulations. Thus it would seem from these observations that the ripples are in great part smoothed away in the pit and that there should be much less destructive action in houses with foundations rising from deep pits than in those built on the free surface*.

Scales of Seismic Intensity

49. Uses of Scales of Intensity. The determination of the maximum acceleration by means of overthrown columns and fractured walls is possible only in strong earthquakes and in their central areas. Nor, on account of their cost, can seismographs be widely used. Various arbitrary scales of seismic intensity have therefore been suggested, their chief objects being (i) to compare the intensities of different earthquakes, and (ii) to trace, by means of isoseismal lines, the variation of intensity in a shock throughout its disturbed area. A good scale may, indeed, be more useful for the latter purpose than accurately constructed seismographs, for it enables us to obtain a large number of observations of the intensity from within a limited area.

50. Conditions to be fulfilled by Scales of Intensity. The following conditions should be fulfilled by any satisfactory scale of intensity:

(i) Each degree should represent a constant intensity. It should depend on the mechanical effects of the shock, and not on personal impressions which may vary in different countries and with different observers in the same country.

(ii) Each degree should consist of one test only, unless the exact equivalence of two or more tests has been determined.

* *Trans. Scis. Soc. Japan*, vol. 16, 1892, pp. 19–45. Observations of a similar, but less detailed, character were made by Milne two or three years earlier (*Trans. Seis. Soc. Japan*, vol. 10, 1887, pp. 25–26, 36). The observations described above show that the deep cuttings through which the Panama Canal passes may to some extent be immune from the effects of local earthquakes.

(iii) The degrees of a scale should be so far apart that an intelligent observer should have no difficulty in distinguishing between the tests of successive degrees; and yet they should be close enough to be applicable to the earthquakes of any country and to shocks of all degrees of strength.

The scales most widely used are the Rossi-Forel scale and the Mercalli scale (sects. 51 and 52). The latter is suitable for strong earthquakes and is adopted in Italy as the standard scale; the former is slightly better adapted to earthquakes of moderate strength and is widely used in other seismic countries.

It will be noticed that neither scale satisfies the first and second of the above conditions. Both depend to some extent on the effect of the shock on the observer, and in each the average number of tests to a degree is three.

51. Rossi-Forel Scale (1883). (1) Recorded by a single seismograph, or by some seismographs of the same pattern, but not by several seismographs of different kinds; the shock felt by an experienced observer.

(2) Recorded by seismographs of different kinds; felt by a small number of persons at rest.

(3) Felt by several persons at rest; strong enough for the duration or direction to be appreciable.

(4) Felt by several persons in motion; disturbance of moveable objects, doors, windows, creaking of floors.

(5) Felt generally by everyone; disturbance of furniture and beds; ringing of some bells.

(6) General awakening of those asleep; general ringing of bells; oscillation of chandeliers, stopping of clocks; visible disturbance of trees and shrubs; some startled persons leave their dwellings.

(7) Overthrow of moveable objects, fall of plaster, ringing of church-bells, general panic, without damage to buildings.

(8) Fall of chimneys, cracks in the walls of buildings.

(9) Partial or total destruction of some buildings.

(10) Great disasters, ruins, disturbance of strata, fissures in the earth's crust, rock-falls from mountains*.

* *Arch. des Sci. phys. et nat.*, vol. 11, 1884, pp. 148–149. A simplified form of this scale (with only one test for each degree), used in the investigation of British earthquakes, is given in *Phil. Mag.*, vol. 50, 1900, p. 51; *Geogr. Journ.*, vol. 46, 1915, pp. 360–361.

52. Mercalli Scale. (1) Instrumental shock, that is, noted by seismic instruments only.

(2) Very slight, felt only by a few persons in conditions of perfect quiet, especially on the upper floors of houses, or by many sensitive and nervous persons.

(3) Slight, felt by several persons, but by few relatively to the number of inhabitants in a given place; said by them to have been *hardly felt*, without causing any alarm, and in general without their recognising it was an earthquake until it was known that others had felt it.

(4) Sensible or moderate, not felt generally, but felt by many persons indoors, though by few on the ground-floor, without causing any alarm, but with shaking of fastenings, crystals, creaking of floors, and slight oscillation of suspended objects.

(5) Rather strong, felt generally indoors, but by few outside, with waking of those asleep, with alarm of some persons, rattling of doors, ringing of bells, rather large oscillation of suspended objects, stopping of clocks.

(6) Strong, felt by everyone indoors, and by many with alarm and flight into the open air; fall of objects in houses, fall of plaster, with some cracks in badly-built houses.

(7) Very strong, felt with general alarm and flight from houses, sensible also out-of-doors; ringing of church-bells, fall of chimney-pots and tiles; cracks in numerous buildings, but generally slight.

(8) Ruinous, felt with great alarm, partial ruin of some houses, and frequent and considerable cracks in others; without loss of life, or only with a few isolated cases of personal injury.

(9) Disastrous, with complete or nearly complete ruin of some houses and serious cracks in many others, so as to render them uninhabitable; a few lives lost in different parts of populous places.

(10) Very disastrous, with ruin of many buildings and great loss of life, cracks in the ground, landslips from mountains, etc.*

53. Absolute Scales of Intensity. The defects of arbitrary scales of intensity are so obvious that some seismologists have attempted to replace them by absolute scales, the maximum intensity in each degree being expressed by the corresponding acceleration in millimetres per second per second.

* *Boll. Soc. Sis. Ital.*, vol. 8, 1902, pp. 184–191.

The earliest scale of this kind is that proposed by Omori, which is applicable only to strong earthquakes, the maximum accelerations corresponding to successive degrees being 300, 900, 1200, 2000, 2500, 4000, and much more than 4000, mm. per sec. per sec. A few years later, Cancani suggested a scale for shocks of all intensities, consisting of twelve degrees, the maximum intensities for each in mm. per sec. per sec. being 2·5, 5,

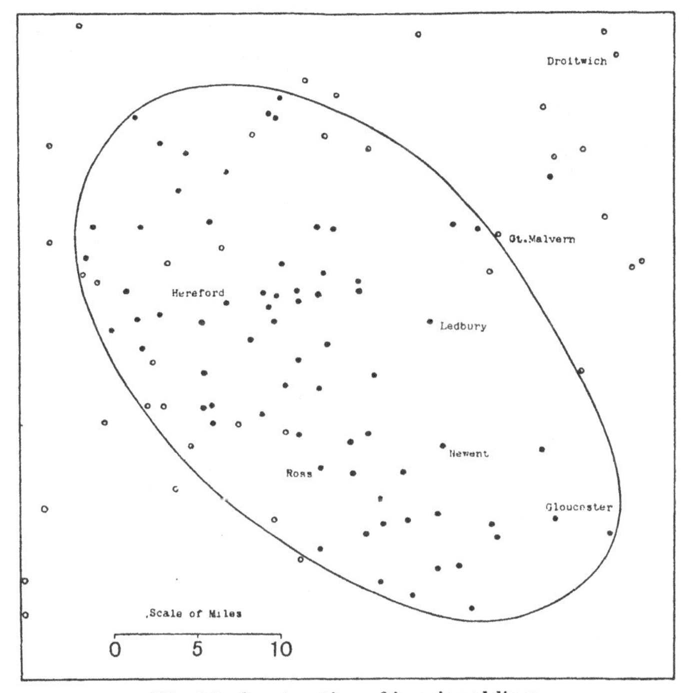

Fig. 23. Construction of isoseismal lines.

10, 25, 50, 100, 250, 500, 1000, 2500, 5000, and 10,000. McAdie has recently brought forward a third absolute scale, which is the same as Cancani's except that the first three degrees of the latter are grouped as one. At the present time, however, we have no means for obtaining numerous determinations of the maximum acceleration when it is less than 200 or 300 mm. per sec. per sec., and thus Omori's scale is the only one that is now of practical value*.

* F. Omori, *Publ. Eq. Inv. Com.*, No. 4, 1900, pp. 14, 137–141; A. Cancani, *Verh. der II. intern. seis. Konf.*, 1904; A. McAdie, *Bull. Seis. Soc. Amer.*, vol. 5, 1915, p. 123.

Isoseismal Lines and Disturbed Area

54. Construction of Isoseismal Lines. The accompanying sketch-map (Fig. 23) illustrates the method of drawing isoseismal lines. It represents the central area, bounded by the isoseismal line of intensity 8, of the Hereford earthquake of

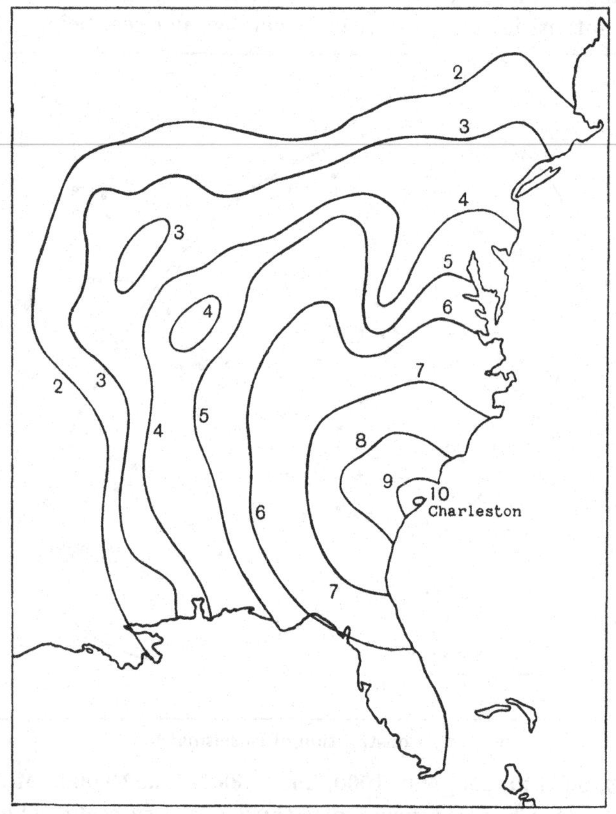

Fig. 24. Isoseismal lines of the Charleston earthquake of 1886.

1896. At all places marked by black dots, the intensity was not less than 8 (Rossi-Forel scale), that is, chimneys were thrown down or walls cracked. Places at which no damage was reported are indicated by small circles. The isoseismal line is then drawn so as to include all the former places and as far as possible to exclude the latter. For a lower degree of intensity, say 6, the observations may provide intensities which are certainly 6, probably 6, possibly 6, and certainly less than 6. The line would then be drawn so as to include places of the first class, many or most of those of the second, some of those of the third, and none of those of the fourth class*.

55. Forms of Isoseismal Lines. Several examples of the forms of isoseismal lines are given in the maps in this volume. As examples of their form in the comparatively slight earthquakes of Great Britain may be mentioned those of the Helston earthquake of 1898 (Fig. 30) and the Derby earthquakes of 1903 and 1904 (Figs. 28 and 51). The isoseismal lines of the Charleston earthquake of 1886 are reproduced in Fig. 24†. Fig. 26 shows the isoseismal lines of the Mino-Owari earthquake of 1891, in which Omori's absolute scale of intensity is used.

The difference in the forms of the Charleston and British isoseismals—the undulations of the one and the regularity of the others—should be noticed. The difference is chiefly due to the use of several tests for each degree of the Rossi-Forel scale in the Charleston earthquake and of one test only for each degree in the British earthquakes; and partly to the very large number of observations available in the British earthquakes. Probably, it is also due in some measure to the wide variations in the nature and form of the ground traversed by the Charleston earth-waves.

56. Magnitude of the Disturbed Area. The disturbed areas of earthquakes range between wide limits. That the area depends to a great extent on the intensity of the shock is clear from the following table, in which is given the disturbed area in British earthquakes (1889–1909) for each degree of the Rossi-Forel scale from 8 to 3:

Intensity	Disturbed area in sq. miles		
	Max.	Min.	Average
8	98000	33000	65900
7	63600	1000	24500
6	3100	74	1200
5	3000	90	850
4	1130	28	260
3	219	81½	126

* For further details regarding the construction of isoseismal lines, see *Beitr. zur Geoph.*, vol. 9, 1908, pp. 214–218.

† Dutton, plate 29.

From the wide difference between the maximum and minimum areas for each degree, it is evident that other factors, besides the intensity, govern the extent of the disturbed area. In volcanic earthquakes, strong enough to ruin the epicentral villages, the area ranges from 50 to about 1000 sq. miles (sect. 227).

As a rule, a destructive tectonic earthquake is felt over an area of one-quarter to one-third of a million square miles, as, for example, over 219,000 sq. miles in the Riviera earthquake of 1887, 230,000 sq. miles in the Bengal earthquake of 1885, about 250,000 sq. miles in the Cachar earthquake of 1869, 330,000 sq. miles in the Mino-Owari earthquake of 1891, and about 373,000 sq. miles in the Californian earthquake of 1906.

Occasionally, much larger areas are shaken. For instance, the Assam earthquake of 1897 disturbed about 1¾ million sq. miles, and the Kangra earthquake of 1905 nearly 2 million sq. miles. The largest known disturbed area is that of the Charleston earthquake of 1886, which covered about 2,800,000 sq. miles. In this case, however, the earthquake was not one of the first order of magnitude, and its extensive disturbed area (bounded by an isoseismal of intensity 2, Fig. 24) was mainly due to its occurrence within an area occupied by civilised races *.

Direction of Earthquake-Motion

57. Direction as revealed by Seismographs. The direction of motion during an earthquake is rarely rectilinear. In successive vibrations, the movement may take place in all azimuths. An example, perhaps an extreme example, of this varied movement is given by Sekiya, who represented the motion of an earth-particle during the Japanese earthquake of Jan. 15, 1887, by a model reproduced from the three components of the motion (see also Fig. 21).

Another example, perhaps also an extreme one, is that of the Tokyo earthquake of June 20, 1894 (Fig. 19). Though the direction of motion, as usual, changed during this earthquake, the maximum horizontal motion was directed towards S. 70° W., and the principal movements both before and after were also in the same or opposite direction. The epicentre was situated to the east of Tokyo.

* *Geol. Mag.*, 1910, p. 412.

The direction of the maximum horizontal motion has been compared with that of the epicentre from the observing station in a number of Japanese earthquakes recorded at Miyako (1896–1898) and Tokyo (1887–1889). When the earthquakes are strong and the epicentres at no great distance, there is in many cases a rough agreement between the two directions. On the other hand, when the earthquakes are weak, there seems to be no prevailing direction discernible in the vibrations*.

58. Mean Direction determined from the Overthrow of Columns. Some interesting observations have been made by Omori on the directions in which columns of various forms were overthrown by the Tokyo earthquake of 1894. In Tokyo, he

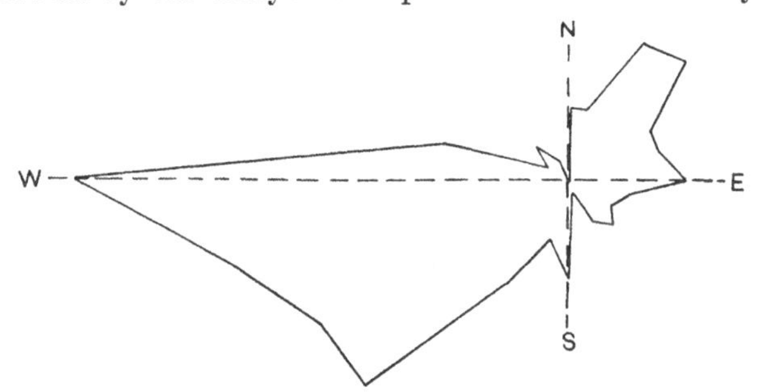

Fig. 25. Direction-rose for the Tokyo earthquake of June 20, 1894.

measured the directions in which 245 ishidoro (stone lanterns in gardens) and other bodies fell. Of these bodies, 144 were ishidoro with circular bases. The columns fell in various directions, the numbers within successive angles of 15° being represented in the diagram in Fig. 25. The mean of the 245 directions was S. 71° W.–N. 71° E., which coincides almost exactly with the direction of the maximum movement recorded in the last section. Of a total number of columns, 159 (or 65 per cent.) fell within the two quadrants adjacent to the direction of maximum movement, and 86 (or 35 per cent.) in the other quadrants.

Similar observations were made by Omori at a number of

* S. Sekiya, *Trans. Seis. Soc. Japan*, vol. 11, 1887, pp. 175–177; S. Sekiya and F. Omori, *Journ. Coll. Sci.*, Imp. Univ. Tokyo, vol. 7, 1894, pp. 1–4; F. Omori, *Publ. Eq. Inv. Com.*, No. 11, 1902, p. 66; F. Omori and K. Hirata, *Journ. Coll. Sci.*, Imp. Univ. Tokyo, vol. 11, 1899, pp. 193–194. Sekiya's model, referred to above, is reproduced in *Nature*, vol. 37, 1888, p. 297.

places within and near the meizoseismal area of the Mino-Owari earthquake of 1891. The area within the dotted line in Fig. 26 represents the zone of extreme violence. The two curves, marked 800 and 2000, are the isoseismal lines referred to in sect. 55. The mean directions of motion in different places are indicated by the short arrows, the arrow-heads pointing in the direction

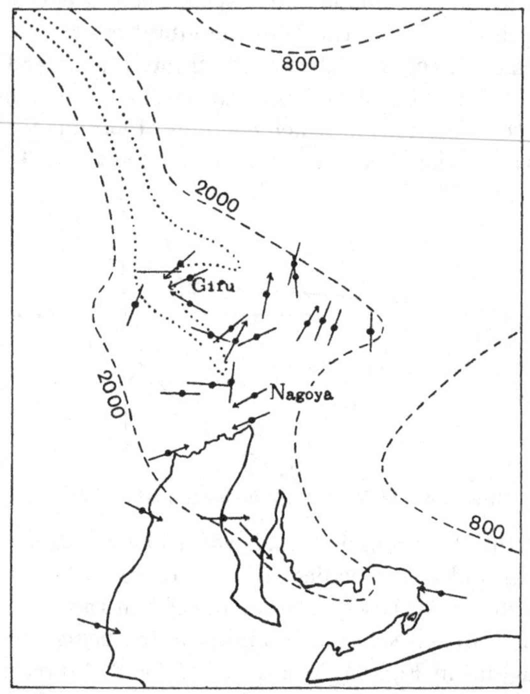

Fig. 26. Map of the directions of the shock and of the absolute isoseismal lines of the Mino-Owari earthquake of 1891.

in which the majority of bodies in each place were overthrown. It will be noticed that, in nearly every case, the mean direction of motion was at right angles to, and directed towards, the meizoseismal zone*.

59. Mean Direction determined by Personal Impressions. Observations on the direction of a shock are greatly influenced by the direction of the principal walls of the house. In four strong

* F. Omori, *Publ. Eq. Inv. Com.*, No. 4, 1900, pp. 17–22, 25–33; *Boll. Soc. Sis. Ital.*, vol. 2, 1896, pp. 184–187.

British earthquakes, the apparent direction of the shock agreed with the direction of the principal walls of the house in 70 per cent. of the observations, the mean deviation of the apparent direction from the direction of the walls being only $9\frac{1}{2}°$. Now, the mean of a large number of observations in any place on the apparent direction is found to coincide very nearly with the direction of the line joining the place to the epicentre. For example, in the Hereford earthquake of 1896, the deviation of the mean apparent direction from the direction of the epicentre was 2° in Birmingham and London; while, in the Derby earthquake of 1904, the two directions were coincident at Derby. The explanation no doubt is that the sense of direction is most apparent in houses in which the principal walls are parallel or perpendicular to the true direction of the shock*.

Duration of Earthquake-Motion

60. Duration of the Preliminary Tremor. With some rare exceptions, the preliminary tremor is present in every seismogram. Its duration, as Omori has shown, depends, not on the strength of the shock, but solely on the distance of the station from the origin. For instance, the preliminary tremor of the Mino-Owari earthquake of 1891 lasted 2 seconds at Gifu, while the average duration in five of the after-shocks was also 2 seconds. On the other hand, the duration for the principal earthquake was 14 seconds at Osaka (distant 87 miles) and 37 seconds at Tokyo (179 miles).

In Fig. 27, the relation between the distance of the origin and the duration of the preliminary tremor is represented for various Japanese earthquakes from 1891 to 1900. The distance (x) is measured in kilometres, the duration (y) in seconds. The corresponding points are grouped close to, and on either side of, the straight line in the figure, the equation of which Omori finds to be

$$x = 7{\cdot}27y + 38.$$

In these earthquakes, the distance of the epicentre lies between 70 and 900 kms. For earthquakes in which the distance lies between 50 and 200 kms., Omori gives the formula

$$x = 6{\cdot}86y + 8{\cdot}1,$$

* C. Davison, *Beitr. zur Geoph.*, vol. 8, 1906, pp. 73–74; vol. 9, 1908, pp. 219–220.

and, for ordinary earthquakes with an origin less than 1000 kms. distant,

$$x = 7{\cdot}42y^*.$$

61. Duration of the Earthquake. To the unaided senses, the total duration of an earthquake varies from one or a few seconds for a weak or moderately strong earthquake to 3 or even 4 minutes for one of destructive violence. In the stronger British earthquakes, the mean duration ranges from 4·0 seconds for the Stafford earthquake of 1916 to 10·5 seconds for the Hereford earthquake of 1896. In great earthquakes, single

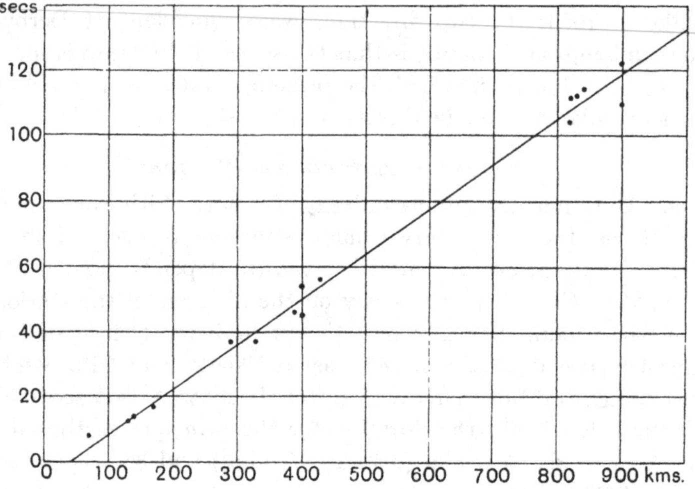

Fig. 27. Relation between the duration of the preliminary tremor and the distance of the station from the origin.

estimates of the duration amount to 70 seconds in the Charleston earthquake of 1886, $3\frac{1}{2}$ minutes in the Californian earthquake of 1906, and as much as 5 minutes in the Assam earthquake of 1897; but, in such cases, it is always possible that the estimates include the duration of closely succeeding after-shocks.

When the movement is registered by seismographs, the duration is much greater. For instance, the records of a Gray-Milne seismograph at Miyako (Japan) from 1896 to 1898, give total durations ranging from 8·5 to 200 seconds. The following table

* F. Omori, *Publ. Eq. Inv. Com.*, No. 13, 1903, pp. 88–91; *Bull. Eq. Inv. Com.*, vol. 1, 1907, pp. 145–154; vol. 2, 1908, pp. 144–147; vol. 6, 1914, p. 238; and vol. 9, 1918, pp. 33–39.

gives the durations of the preliminary tremor, principal portion and end-portion in some of these earthquakes, the principal portion being clearly defined only when the epicentre is not more than 200 kms. distant.

Earthquake	Distance of epicentre in kms.	Duration in seconds		
		Preliminary tremor	Principal portion	End-portion
1896. Aug. 31 (1)	100	10	4	59
,, (2)	100	5·2	26	69
1897. Mar. 27	70	4·2	24	52
Apr. 30	60	2·6	0·7	39
June 18	90	8·9	9	32
Aug. 23	110	11	9·3	80
Oct. 2	150	12·5	13·6	84
Dec. 23	70	3·6	3·6	28
1898. Apr. 23	200	13	1·9	105
Average	—	7·9	10·2	61

Of 487 earthquakes recorded in 1905 by an Omori horizontal tremor recorder at Mount Tsukuba (Japan), the total durations varied from 5 to 275 minutes, 64 per cent. of the total number having durations of less than 1 minute, 28 per cent. between 1 and 2 minutes, 4 per cent. between 2 and 3 minutes, 2 per cent. more than 3 minutes, the durations of the rest being indeterminate.

Again, the apparent duration of an earthquake depends to a great extent on the sensitiveness of the seismograph employed. For instance, the duration of one of the above earthquakes was 2 minutes according to the record of the Gray-Milne seismograph at Miyako, and 2 hours as registered by an Omori horizontal pendulum at Tokyo; the former instrument responding to the quicker vibrations only and not to the slow undulations forming the end-portion of the movement*.

* F. Omori and K. Hirata, *Journ. Coll. Sci.*, Imp. Univ. Tokyo, vol. 11, 1899, pp. 189–191 and table; F. Omori, *Publ. Eq. Inv. Com.*, No. 22 A, 1908, pp. 1–39.

CHAPTER IV

THE SOUND-PHENOMENA OF EARTHQUAKES

62. The vibrations which are perceptible as sound have a wide range of periods, the lowest audible note being produced by about 30 vibrations per second, and the highest by about 70,000. Both limits are, however, subject to variation, in different persons as well as in different races*.

63. General Character of the Sound. The sound which accompanies an earthquake is usually a low heavy rumbling noise, a deep booming or low moaning, a grating roaring or a crushing grinding noise. Sometimes, very different sounds are heard like that of a roaring wind or of a chimney on fire.

The chief characteristic of the sound is its extraordinary depth. It is almost too low to be heard. It is described as a rumble that can be felt. The impression of great depth is also conveyed by the frequent use of the word "heavy," in such comparisons as heavy peals of thunder, heavy gusts of wind, or the heavy rumbling of sea-waves in a cave. But no evidence of its extreme lowness is so decisive as the fact that the sound is heard by some observers and not by others. To one the sound

* The following papers may be consulted, on the general phenomena of earthquake-sounds:

1. Davison, C. (1). On earthquake-sounds. *Phil. Mag.*, vol. 49, 1900, pp. 31–70.

2. —— (2). The sound-phenomena of British earthquakes. *Beitr. zur Geoph.*, vol. 12, 1913, pp. 485–527;

and on brontides:

3. Alippi, T. (1). I mist-poeffeurs calabresi. *Boll. Soc. Sis. Ital.*, vol. 7, 1901, pp. 9–22.

4. —— (2). Di un fenomeno acustico della terra o dell' atmosfera. *Ibid.* vol. 12, 1907, pp. 1–42.

5. —— (3). Nuovo contributo all' inchiesta sui "Brontidi." *Ibid.* vol. 15, 1911, pp. 65–77.

6. Cancani, A. (1). Barisal-guns, mist-poeffeurs, marina. *Ibid.* vol. 3, 1897, pp. 222–234.

7. —— (2). Rombi sismici. *Ibid.* vol. 7, 1901, pp. 23–47.

8. Van den Broeck, E. Un phénomène mystérieux de la physique du globe. *Ciel et Terre*, vol. 16, 1895, and vol. 17, 1896.

seems like that of a loud explosion, to another in the same place like distant thunder, to a third the shock appears unaccompanied by sound.

64. Types of Earthquake-Sound. Occasionally, the sound which accompanies an earthquake is supposed to be unlike any known sound. As a rule, however, it is compared to one of the types of the following scale, known as the Davison sound-scale:

1. Waggons, carriages, traction-engines or trains passing, generally very rapidly, on hard ground, over a bridge or through a tunnel; the dragging of heavy boxes or furniture over the floor.
2. Thunder, a loud clap or heavy peal, but most often distant thunder.
3. Wind, a moaning, roaring or rough strong wind; the rising of the wind, a heavy wind pressing against the house, the howling of wind in a gap or a chimney, a chimney on fire, etc.
4. Loads of stones, etc., falling, such as the tipping of a load of coals or bricks.
5. Fall of heavy bodies, the banging of a door, the blow of a wave on the seashore.
6. Explosions, distant blasting, the boom of a distant heavy gun.
7. Miscellaneous, such as the trampling of many animals, an immense covey of partridges on the wing, the roar of a waterfall, a low pedal note on the organ, and the rending or settling together of huge masses of rock.

In strong British earthquakes (those in which the isoseismal 4 includes an area of more than 5000 sq. miles), 46 per cent. of the observers compare the sound to passing waggons, etc., 24 per cent. to thunder, 11 per cent. to wind, 5 per cent. to loads of stones falling, 3 per cent. to the fall of a heavy body, 7 per cent. to explosions, and 5 per cent. to miscellaneous sounds. In very slight earthquakes (disturbing areas of less than 60 sq. miles), the figures are different: 9 per cent. compare the sound to passing waggons, etc., 11 per cent. to thunder, 2 per cent. to wind, 9 per cent. to loads of stones falling, 25 per cent. to the fall of a heavy body, 42 per cent. to explosions, and 2 per cent. to miscellaneous sounds.

As a rule, the sound adheres throughout to one of the types mentioned above, and varies, if at all, only in intensity, becoming gradually louder and then dying away. The places at which changes of type are observed are situated for the most part within a district closely surrounding the epicentre, and the changes take place at the moment when the sound is loudest and the strongest vibrations are felt. Occasionally, a loud explosive crash, like that of heavy blasting, is heard at this moment, or a sound like that of rushing wind may merge into that of a heavily loaded traction-engine passing at a rapid rate*.

65. Inaudibility of the Sound. The inaudibility of the sound is either total or partial. Partial inaudibility may consist in the suspension of all sound during part only of the time while it is heard by others, or in the suppression of some vibrations only, so that observers in the same place may refer the sound to different types† or that some may hear the deep explosive crashes of which others are unconscious.

In British earthquakes, the sound is heard almost invariably, the omission of reference to the sound in exceptional cases being probably accidental. In strong earthquakes, 83 per cent. of all the observers hear the sound. In slight earthquakes (those in which the isoseismal 4 includes areas of less than 1000 sq. miles), the percentage rises to 97. For this higher percentage, there are two reasons: (i) the sound in slight earthquakes is usually a more prominent feature than the shock, and (ii) in strong earthquakes, the sound-area is less extensive than the disturbed area, while in slight earthquakes the two areas are approximately coincident.

The inaudibility of the sound to some observers is often attributed to inattention. When the sound appears to some as a deafening noise or as louder than the loudest thunder, such an explanation must obviously fail. As this is the case, however, near the epicentre only, a more decisive test is furnished, in the case of earthquakes which occur at night, by the audibility of that part of the sound before the vibrations begin to be felt. In the Hereford earthquake of 1896 (5.32 a.m.), the

* Davison (2), pp. 488–497.

† For instance, at Birmingham during the Hereford earthquake of 1896, 36 per cent. of the observers refer to passing waggons, 18 per cent. to thunder, 18 per cent. to wind, 4 per cent. to loads of stones falling, 6 per cent. to the fall of a heavy body, 11 per cent. to explosions, and 7 per cent. to miscellaneous sounds.

fore-sound was heard by 72 per cent. of those who were awake and 74 per cent. of those who were asleep; in the Inverness earthquake of 1901 (1.24 a.m.), the corresponding figures are 72 and 72; and in the Doncaster earthquake of 1905 (1.37 a.m.) 78 and 67. Moreover, in the Hereford earthquake, the percentages within the isoseismal 8 were 78 for those awake and 75 for those asleep; at a considerable distance, in the zone between the isoseismals 6 and 5, the corresponding figures were 61 and 60. Thus, roughly, the preliminary sound is equally audible to all observers, whether awake or asleep, and therefore the inaudibility of the sound to some cannot be due to inattention. It can only be explained on the supposition that the sound-vibrations of an earthquake are in the immediate neighbourhood of the lower limit of audibility, and that this limit varies in different persons, so that some may be deaf to such low sounds though by no means deaf to ordinary noises.

The partial inaudibility of the sound is no doubt due in part to the same cause, especially as regards the deep explosive crashes heard by some, and not by others, when the shock is strongest. The temporary cessation (to some) of all sound while the shock is felt may be due to this cause, partly also to fatigue*.

66. Variation in Audibility throughout the Sound-Area. In strong British earthquakes, it is usually possible to calculate the percentage of audibility in five zones bounded by successive pairs of isoseismals. The average percentages for such earthquakes are 97 within the central isoseismal, and, in successive zones, 94, 88, 69 and 60. There is thus, as we should expect, a decline in audibility as the distance from the origin increases, but the rate of decline is at first slow, and afterwards more rapid, especially near the boundary of the sound-area. From this rapid decline, we may infer that the lower limit of audibility does not vary much in different observers†.

67. Isacoustic Lines. Isacoustic lines are lines which pass through all points at which the percentage of audibility of the earthquake-sound is the same.

For their construction, the sound-area is divided into equal areas, and the percentage of audibility within each is supposed

* Davison (1), pp. 39–43; Rayleigh, *Nature*, vol. 56, 1897, p. 285.

† Davison (2), pp. 505–506.

to be equal to that at its centre. Curves corresponding to different percentages are then drawn through the points at which such percentages are affixed, or through points which divide the lines joining successive centres in the proper proportion. Fig. 28 shows the isacoustic lines corresponding to percentages 95 and 90 for the Derby earthquake of 1903 (see also Fig. 51). The significance of these lines will be referred to in a later section dealing with twin earthquakes (sect. 242)*.

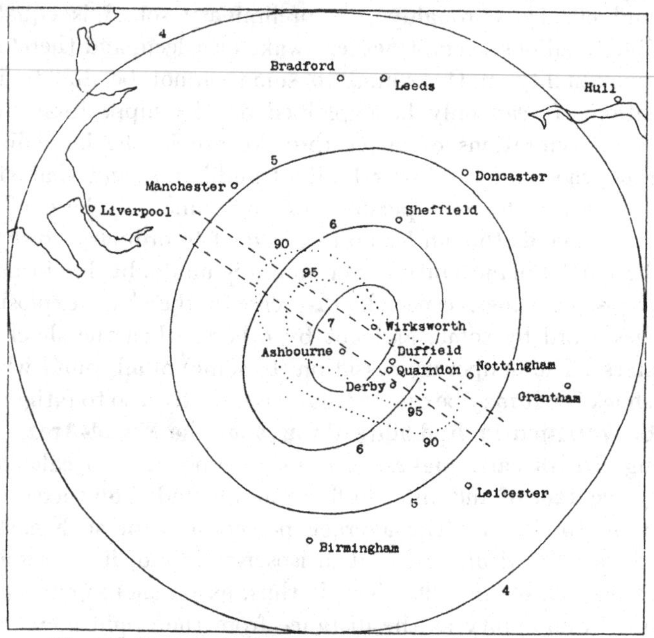

Fig. 28. Map of the Derby earthquake of Mar. 24, 1903.

68. Variation in Character throughout the Sound-Area. Throughout the sound-area, the sound maintains its uniform lowness of pitch, if we may judge from the frequency with which the word "heavy" is used in descriptions of the sound. As the distance from the epicentre increases, there is a steady decline in the percentage of references to thunder, loads of stones falling and explosions, a steady increase in the references to wind, and on the whole an increase in the frequency of comparisons to passing waggons. Among the comparisons to

* Davison (2), pp. 506–508.

thunder, there is a steady and rapid increase in the percentage of references to distant thunder. Thus, there is a greater monotony, an approach to uniformity both in intensity and pitch, as the distance from the epicentre increases*.

69. Relative Magnitude of Sound-Area and Disturbed Area. The relative magnitude of the sound-area and disturbed area ranges continuously between the widest limits. On the one hand, the shock is felt but is unaccompanied by sound; on the other hand, the sound is heard without any attendant shock.

In the great majority of strong and violent earthquakes, the sound-area occupies a region surrounding the epicentre, while the disturbed area extends beyond it in every direction. For instance, in the Verny (Turkestan) earthquake of 1887, the disturbed area contained about 400,000 sq. miles, the sound-area about 132,000 sq. miles. In the Italian earthquake of 1873, the two areas contained respectively 227,000 and 22,000 sq. miles. In Japan, 30 per cent. of the earthquakes during 1885–1892 which disturbed areas of more than 10,000 sq. miles were unaccompanied by recorded sound; and, when heard, the sound as a rule was inaudible at more than a few miles from the epicentre. Of the earthquakes which originated beneath the land during these years, 26·5 per cent. are recorded as accompanied by sound. For the submarine earthquakes of the same period, the corresponding percentage is 0·84. None of the earthquakes originated at a greater distance than 40 or 50 miles from the shore, while the epicentres of 93 per cent. of the total number were not more than 10 miles distant.

In strong British earthquakes, the disturbed area extends beyond, but not far beyond, the sound-area in all directions. For instance, in the Hereford earthquake of 1896, the disturbed area contained 98,000 sq. miles, and the sound-area 70,000 sq. miles. On an average, the sound-area is about two-thirds of the disturbed area. In British earthquakes of moderate strength, and in some slight earthquakes, the sound-area and disturbed area practically coincide. The areas in such cases range from 400 to 2000 sq. miles, and in one case to 4340 sq. miles. In many slight British earthquakes, the sound-area overlaps the disturbed area, as a rule on one side only, sometimes in every direction. Examples of the partial overlapping will be given in sect. 72.

* Davison (2), pp. 508–513.

Still lower in the scale are earth-sounds, in which sounds alone are observed without the slightest accompanying tremor. As a general rule, they appear to form part of the series of after-shocks of a great earthquake, or occur as intercalated members of a series of weak shocks. For instance, 3365 after-shocks of the Mino-Owari earthquake were recorded at Gifu from Oct. 28, 1891, to the end of 1893, and of these 409 were earth-sounds. After the Comrie (Perthshire) earthquake of Oct. 23, 1839, one observer at Comrie noted, between this date and the end of 1841, 44 shocks and 234 earth-sounds.

The earth-sounds in this latter district are of considerable interest. They are heard in an area in which slight shocks, accompanied by precisely similar sounds, are at times very frequent. Here there is a complete continuity from earthquake to earth-sound; from the strong earthquake in which the disturbed area extends in all directions beyond the sound-area, through the moderate earthquake in which both areas coincide approximately, and the slight earthquake in which the sound-area overlaps the disturbed area in one or every direction, down to the earth-sound when the disturbed area vanishes*.

70. Earth-Sounds. In addition to the examples given in the preceding section, reference may be made to three cases in which earth-sounds were especially numerous.

In the island of Meleda, in the Adriatic Sea, earth-sounds were frequently heard during the years 1822–1826. Partsch gives a list of the shocks and sounds observed from Nov. 17, 1824, to Feb. 18, 1826. In this interval, there were 30 shocks and 71 detonations. Of the shocks, all but three were accompanied by sound.

Another district in which earth-sounds were at one time frequent is that surrounding East Haddam in Connecticut. Before the English settlements, the sounds were well known to the Indian inhabitants, who called the place Morehemoodus or place of noises. According to an observer, writing in 1729, eight or ten sounds, resembling small arms, were sometimes heard in 5 minutes, and great numbers in the course of a year. Often they could be heard "coming down from the north imitating

* Davison (1), pp. 53–56; (2), pp. 513–515; F. Omori, *Journ. Coll. Sci.*, Imp. Univ. Tokyo, vol. 7, 1894, p. 113; J. Drummond, *Phil. Mag.*, vol. 20, 1842, pp. 240–247.

slow thunder, until the sound came near, or right under, and then there seemed to be a breaking, like the noise of a cannon-shot, or severe thunder, which shakes the houses and all that is in them." At the end of the eighteenth century, they were still heard frequently. At the present time, they seem to have ceased.

Lastly, Humboldt describes a remarkable series of earth-sounds as the subterranean thunder of Guanaxuato, a city on the Mexican plateau, far removed from any active volcano. From Jan. 13–16, 1784, "it seemed to the inhabitants as if heavy clouds lay beneath their feet, from which issued alternate slow rolling sounds and short quick claps of thunder." In this case, the earth-sounds were not accompanied by sensible shocks *.

We may infer from these and other examples: (i) that earth-sounds are heard generally in those districts in which slight shocks are frequent; (ii) that, in the midst of a series of earth-sounds, slight shocks, accompanied by precisely similar sounds, are occasionally intercalated, there being a complete continuity from earthquake to earth-sound. We may therefore conclude that earthquakes and earth-sounds are manifestations, differing only in degree and in the method in which we perceive them, of one and the same phenomenon.

71. Brontides. In many parts of the world, there are heard sounds closely resembling those described in the preceding paragraphs. In the delta of the Ganges they are known as Barisal-guns, on the coasts of Belgium as mist-poeffeurs, in central Italy as marinas, in Haiti as gouffres. They have been called brontides by Alippi, who has paid special attention to the subject.

The sounds are invariably deep and last as a rule for 5 or 6 seconds; they begin feebly, grow rapidly in strength, and then as rapidly die away. They bear a close resemblance to the booming of guns or distant thunder, though other types of the Davison scale are referred to. In Italy, for instance, 11 per cent. of the descriptions collected by Alippi are referred to passing waggons, 41 per cent. to thunder, 10 per cent. to wind, 6 per cent.

* P. Partsch, *Bericht über das Detonations-Phänomen auf der Insel Meleda bey Ragusa* (Wien, 1826), pp. 204–211; W. T. Brigham, *Mem. Boston Soc. Nat. Hist.*, vol. 2, 1871, pp. 14–16; A. von Humboldt, *Cosmos* (Bohn's edition), vol. 1, pp. 203, 205–206.

to loads of stones falling, 3 per cent. to the fall of a heavy body, 21 per cent. to explosions, and 8 per cent. to miscellaneous sounds. Thus, as regards type of sound, brontides bear some resemblance to the sounds which accompany slight earthquakes. That the sound of brontides is close to the lower limit of audibility is evident from the fact, mentioned by Cancani, that they are heard by some and not by others placed in the same conditions.

Most frequently, single detonations are heard, but they sometimes occur in groups. According to 66 per cent. of the observers consulted by Alippi, the sound appears to travel through the air, according to 25 per cent. through the ground, and according to 9 per cent. through both air and ground.

They are not heard with any approach to uniformity throughout any country. Alippi has mapped the districts in Italy that are subject to brontides, and his map shows that they are confined to special regions, two or three embracing a large part of a province, the majority small. Certain districts, such as the western Alps, are quite free from brontides.

In Italy, brontides are heard with nearly equal frequency in all seasons of the year, though rather more frequently in summer; in the Philippine Islands, they are most frequent in the hot season (March, April and May). They occur most often about sunset and sunrise, least often at night; but this concentration at certain hours may be apparent only, and depending on the habits of the observers.

Notwithstanding the attention that has been paid to brontides, their cause is still somewhat obscure. Cancani urges that they cannot be due to stormy seas, for they are observed often with a calm sea and at considerable distances from the coast. Nor can they be caused by gusts of wind rushing through mountain gorges, for they are heard indifferently on the summits of mountains, on the coast and in open plains, and most frequently when the air is still. It seems improbable that their origin is connected with the atmosphere, for, in that case, they should be heard everywhere, and not in special regions only. An artificial origin is excluded, for they are observed at times when guns and mines are not fired, and in places (such as some African countries) where explosives are unknown.

Cancani thus concludes in favour of a seismic origin for brontides. In support of this, he urges that brontides predominate in countries which are subject to earthquakes, that they are often heard as heralds of earthquakes, and are specially frequent during seismic series, and that brontides are sometimes accompanied by very feeble tremors. Again, in the south-west of Haiti, brontides are very common, especially in the mountain range of La Selle. On the north side, this range is bounded by a steep cliff formed by displacements along a fault that is believed to be still growing. The sounds appear to come from the base of this cliff, and, as they do not differ from those which accompany sensible earthquakes, it is probable that they are caused by small readjustments of the crust along this fault.

On the other hand, in the Philippine Islands, according to Saderra Masò, brontides have little apparent connexion with earthquakes. The Belgian coasts are rarely, if ever, visited by local earthquakes. Moreover, Alippi has compared his brontide map of Italy with Baratta's seismic map of the same country, and he finds that, while the areas in which brontides are common are in many cases (such as southern Calabria) the same as those in which earthquakes are numerous and strong, yet that some of the brontide areas are free from earthquakes, while brontides are unknown in some seismic areas.

It seems difficult, therefore, to avoid the conclusion that brontides have more than one origin, but that in many cases they must be regarded as the same in nature and origin as the earth-sounds described above. It should also be remembered that they are not of necessity the results of recent earth-movements. It is possible that they may be the latest representatives of a series of after-shocks of some long-past and almost forgotten earthquakes*.

72. Relative Position of Sound-Area and Disturbed Area. The excentricity of the sound-area with respect to the isoseismal lines is one of the most significant phenomena of earthquake-sounds. In the strong earthquakes of this country, the isacoustic lines show a remarkable independence of the isoseismal lines. It is in the moderate and slight earthquakes, however,

* Alippi (2), pp. 21–41; Cancani (1), pp. 231–234; J. Scherer, *Bull. Seis. Soc. Amer.*, vol. 2, 1912, pp. 230–232.

that the excentricity is manifested most clearly by the sound-area overlapping an isoseismal line or the boundary of the disturbed area in one direction.

For instance, in the Bolton earthquake of 1889, the isoseismal lines (indicated by the continuous lines in Fig. 29) are nearly circular, the boundary of the sound-area (indicated by the dotted

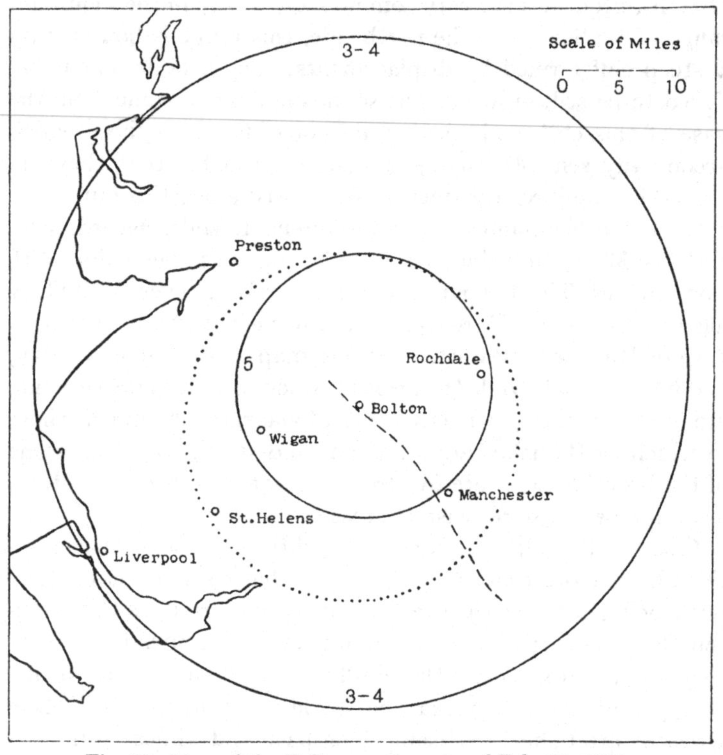

Fig. 29. Map of the Bolton earthquake of Feb. 10, 1889.

line) is also nearly circular, but its centre lies $3\frac{1}{4}$ miles south-south-west of the centre of the isoseismal 5. The great Irwell Valley fault, the course of which is represented on the map by the broken line, hades to the north-east and therefore beneath the epicentre. As the earthquake was probably caused by a slip along this fault in the neighbourhood of Bolton, it follows that the sound-area with respect to the isoseismal 5 is shifted towards the fault-line.

Again, in Fig. 30, the continuous lines represent the isoseismals 3 and 4 of the Helston earthquake of 1898, and these show that the originating fault must hade to the south-east (sect. 129). The outer dotted line indicates the boundary of the sound-area; and the inner dotted line, which is concentric with the other, separates the places where the sound was very loud from those where it was distinctly fainter. In this case, also, the sound-area

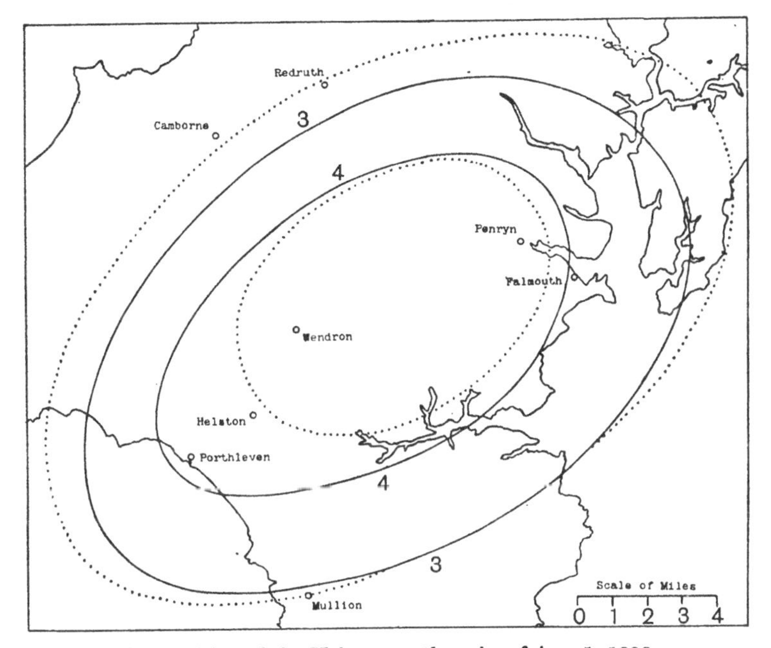

Fig. 30. Map of the Helston earthquake of Apr. 1, 1898.

relatively to the disturbed area is displaced towards the fault-line*.

73. Time-Relations of the Sound and Shock. The earthquake-sound almost invariably accompanies the shock. In British earthquakes, it usually also precedes and follows the shock, overlapping it by one or a few seconds at both ends. This overlapping persists in all parts of the sound-area, for, in the four zones bounded by successive isoseismals, the beginning of the

* C. Davison, *Geol. Mag.*, 1891, pp. 306–316; *Quart. Journ. Geol. Soc.*, vol. 56, 1900, pp. 1–7.

sound precedes that of the shock in 67, 69, 70 and 62 per cent. of the records respectively; while the end of the sound follows that of the shock in 38, 44, 41 and 36 per cent. in the same zones. It would seem, then, that the precedence of the sound is due to a difference in the place of origin rather than to the sound-waves travelling with a greater velocity than the large waves*.

* Davison (2), pp. 517–522.

CHAPTER V

DEFORMATIONS OF THE EARTH'S CRUST

74. The deformations of the crust observed with some great earthquakes take the form of (i) fault-displacements, and (ii) warping of the surface-beds. They are thus of the same character as those which occur in lower layers of the crust, but, in the more rigid outer crust, faulting, as might be expected, predominates over warping*.

The connexion between the crustal deformations and the earthquakes is shown by the coincidence between their times of occurrence and the areas affected by them. It is important, however, to notice that the deformations are not consequences of the earthquakes, but rather, as will be seen in Chapter XIV, primary causes of the earthquakes.

The number of earthquakes accompanied by crustal deformations is considerable. In many cases, however, while it is needless to doubt the reality of the permanent movements, the observa-

* The more important memoirs in which deformations of the earth's crust are described are the following:

1. Fuller, M. L. The New Madrid earthquake. *Bull. U. S. Geol. Surv.*, No. 494, 1912, pp. 1–112.
2. Hobbs, W. H. The earthquake of 1872 in the Owens Valley, California. *Beitr. zur Geoph.*, vol. 10, 1910, pp. 352–385 (especially pp. 371–384).
3. Koto, B. The cause of the great earthquake in Central Japan, 1891. *Journ. Coll. Sci.*, Imp. Univ. Tokyo, vol. 5, 1893, pp. 295–353.
4. Lawson, A. C. (editor). *The Californian Earthquake of April* 18, 1906, vol. 1 and atlas, 1908; vol. 2 (by H. F. Reid), 1910.
5. Lyell, C. *Principles of Geology*, 12th edit., 1875, pp. 82–89.
6. Oldham, R. D. Report on the great earthquake of 12th June, 1897. *Mem. Geol. Surv. India*, vol. 29, 1899, pp. 1–379 (especially pp. 138–163).
7. Omori, F. Preliminary note on the Formosa earthquake of March 17, 1906. *Bull. Eq. Inv. Com.*, vol. 1, 1907, pp. 53–69 (see also pp. 70–72).
8. Tarr, R. S., and L. Martin. The earthquakes at Yakutat Bay, Alaska, in September 1899. *U. S. Geol. Surv.*, Prof. Paper No. 69, 1912, pp. 1–135 (especially pp. 18–45).
9. *Relazione della Commissione Reale incaricata di designare le zone più adatti per la reconstruzione degli abitati colpiti dal terremoto del* 28 *dicembre* 1908, *ecc.* Roma, 1909, pp. 131–156.

tions recorded add little to our knowledge beyond the fact that some movement, usually of elevation, has taken place. Eliminating all such earthquakes, there remain fourteen in which the accompanying displacements have been observed and described with some care. These earthquakes are those of: (1) New Madrid (U.S.A.) in 1811–12; (2) Wellington (N.Z.) in 1855; (3) Owens Valley (U.S.A.) in 1872; (4) Sonora (U.S.A.) in 1887; (5) Mino-Owari (Japan) in 1891; (6) Sumatra in 1892; (7) Baluchistan in 1892; (8) Locris (N.E. Greece) in 1894; (9) Assam in 1897; (10) Alaska in 1899; (11) Kangra (India) in 1905; (12) Formosa in 1906; (13) California in 1906; and (14) Messina in 1908. The displacements of the first and third of these earthquakes left traces which are still visible; in the other cases, they were observed and measured as a rule soon after their formation.

Fault-Displacements

75. The principal features of the fault-movements which occur during earthquakes are the great length of the fault over which the movement takes place and the general uniformity in the direction of the fault-line. The displacement may be almost entirely horizontal or almost entirely vertical, but in most cases both horizontal and vertical. Horizontal displacements are usually manifested by the relative shifting of objects previously in contact or in line; vertical displacements by the formation of fault-scarps. When the displacements are of considerable magnitude, a new trigonometrical survey of the central district, if one has been made not many years before, may convert relative into absolute measurements.

One other feature may be referred to at this stage, namely, that the faulting occurs repeatedly along the same line. The Baluchistan fault has the appearance of an old road, and the natives declare that the ground always cracks along the old fault-line with every severe earthquake. The Owens Valley contains many scarps formed before 1872, the height of one of which was doubled in that year. The southern part of the San Andreas fault in California lies in the desert part of the Coast Ranges, in which erosion takes place slowly. With every great earthquake, the fissure opens anew; so that, to the inhabitants of the district, the fault is known as the "earthquake-crack."

Of the fourteen earthquakes referred to in sect. 74, the deformation in all but three (those of New Madrid, Kangra and Messina) took the form of faulting. In the exceptional cases, the only phenomenon observed was that of warping, either local or general, but it is not impossible that this warping was merely the surface equivalent of deep-seated faulting. In two other

Fig. 31. Fault-scarp of the Mino-Owari earthquake of 1891.

earthquakes (Assam and Alaska), both faulting and warping were present, and this seems to lend some support to the connexion.

76. General Appearance. The superficial effects of the faulting vary with the direction of the movement and also with the nature of the ground traversed. In the case of the Sumatra earthquake of 1892, there was no actual trace of the fault at the surface, and its existence was only revealed by geodetic measurements. In all the other earthquakes, one or more of the faults concerned could be followed for some distance.

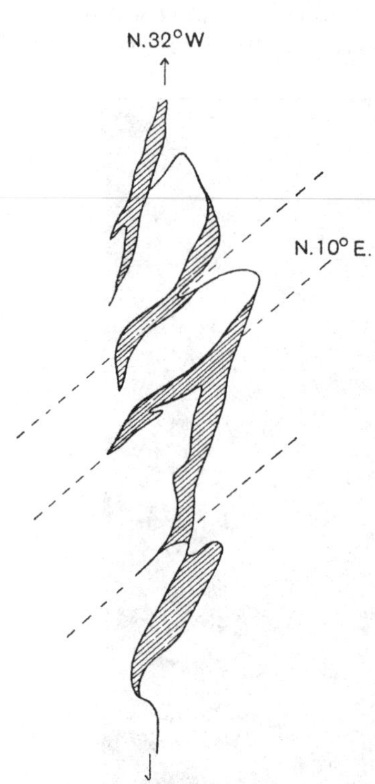

Fig. 32. Secondary cracks in the Californian earthquake of 1906.

When the displacement is horizontal or nearly so, the fault appears as a fine crack in hard rock, as in parts of the Mino-Owari fault, or as an open fissure in earthy ground, as in the Locris fault. In such cases, the relative shifting of objects on either side of the fault is the clearest evidence of the displacement.

When the displacement is partly or wholly vertical, the superficial effects are more distinct. If the surface consist of soft earth and if the uplift be small, the fault appears like a rounded ridge, from 5 to 10 feet wide and about 2 feet in height, as if the soil had been raised by a gigantic mole creeping underground (Fig. 31). In the Californian earthquake of 1906, the formation of this ridge was due partly to the shearing action and partly to compression along the line of fault, and was in many places accompanied by a series of secondary cracks extending a few hundred feet from the fault. Some of these are shown in Fig. 32, the broken lines representing the directions of the secondary cracks which are inclined to the direction of the fault-line at an angle of 42°.

Uplifts of more than 2 feet in amount result in the formation of fault-scarps. In compact rock, the scarps appear as long cliffs, which in the cases known, are always vertical (Fig. 33). In alluvial or earthy ground, the scarps soon weather down into

Fig. 33. Minor fault, Alaskan earthquake, 1899.

Fig. 34. Fault-scarp at Midori of the Mino-Owari earthquake of 1891.

a uniform slope, so that from a distance they resemble railway-embankments (Fig. 34). If, however, the alluvium be of great thickness, the displacement may fail to reach the surface, and the scarp is then replaced by a slope, that is, by warping. For instance, the Chedrang fault, formed during the Assam earthquake of 1897, runs at its north end (Fig. 38) beneath a thick mass of alluvium, converting what was level ground into a smooth unbroken slope, on which trees are tilted over, as shown in Fig. 35.

77. Length. The length of the fault-displacement varies between wide limits. It is least when the fault-system is complex, greatest when the displacement occurs along a single fault.

Fig. 35. Slope in alluvium over the Chedrang fault.

Thus, in the Assam earthquake of 1897, there were two principal faults, the Chedrang fault not less than 12 miles in length, and the Samin fault about 2½ miles long. In the Alaskan earthquakes of 1899, the longer faults (*A*, *E* and *B*, Fig. 39), inferred from the evidence of the vertical displacements, are 18, 16 and 13 miles in length respectively, though it is possible that the first and third form a single fault 31 miles long.

In the following earthquakes, the movement at the surface was practically confined to a single fault in each case; though possibly, as in the Mino-Owari earthquake of 1891, more than one deep-seated fault may have been in action. In the Baluchistan earthquake of 1892, the movement affected a distance of several miles, and the fault itself has been traced for 120 miles. The Formosa fault (1906) was also traced for only part of its length; the entire displacement is, however, estimated by Omori at about 30 miles. The Locris fault (1894) was 34 miles long, the total length of the Owens Valley fault-system (1872) about 40 miles. The Mino-Owari fault (1891) was traced for 40 miles, but there can be little doubt that its length at the surface was 70 miles. The Wellington fault-scarp (1855) was a mainly vertical cliff 90 miles in length. The fault connected with the Sumatra earthquake of 1892 was detected only by the re-triangulation of the district; its length was certainly much greater than 34 miles, and, according to Reid, may have been as much as 125 miles. No more remarkable displacement is known to have occurred than that which accompanied the Californian earthquake of 1906. Though lying for about 70 miles under the sea, the total length was almost certainly 290 miles*.

The displacement during any given earthquake does not necessarily occur over the whole length of the fault concerned. On this point, our information is limited to the faults of the Owens Valley and Californian earthquakes, the latter of which has been studied in greater detail than any other. This, the San Andreas fault, has been traced, with three interruptions in which its course is submarine, from near Cape Mendocino on the north, past San Francisco, to the north end of the Colorado Desert on the south, that is, for a distance of more than 600 miles. During the earthquake of 1906, the movement was confined to its northern half, from near Cape Mendocino to San Juan, 82 miles south-east of San Francisco†.

78. Form. The form of the faults with which earthquakes are connected is similar to that of the faults determined by other evidence. Small faults, such as the Chedrang and Samin faults of the Assam earthquake, are practically straight. Longer faults, such as those in Baluchistan and Locris, maintain a nearly uniform direction through their entire known courses. The

* Oldham, pp. 146–148; Tarr and Martin, p. 35 and plate 14; C. A. and A. H. McMahon, *Quart. Journ. Geol. Soc.*, vol. 53, 1897, p. 292; Omori, pp. 58–61; S. A. Papavasiliou, *Compt. Rend. Acad. Sci. Paris*, vol. 119, 1894, p. 114; Hobbs, pp. 378–381; Koto, p. 349; Lyell, p. 86; Reid, pp. 76–77; Lawson, vol. 1, p. 54.

† Lawson, vol. 1, pp. 48, 54.

direction of the longest of all, the San Andreas fault, between San Juan and Point Arena varies only from 30° to 40° W. of N. (Fig. 36). When drawn on a map of small scale, such as that of Fig. 36, the fault as a whole appears as a nearly even line, slightly curved and convex to the Pacific. On a large-scale map, however, it is seen that the fault is not a smooth uniform curve, but a succession of slightly curved rather than straight portions, the curvatures varying in direction.

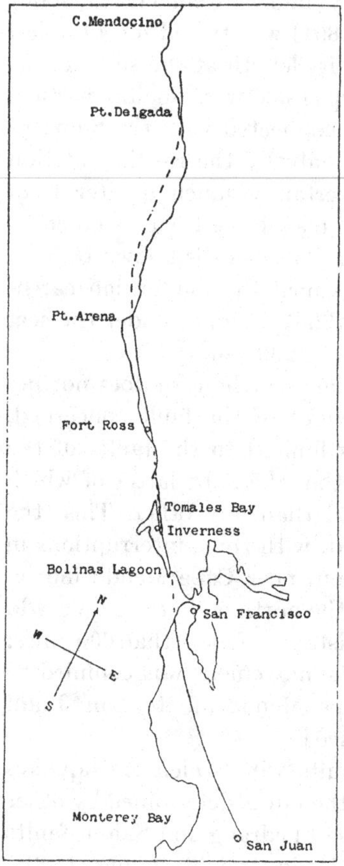

Fig. 36.
Map of the San Andreas fault.

In a few cases, the faults are not single. Though only one great scarp was observed with the Mino-Owari earthquake of 1891, there must have been a second and more deeply-seated fault along which most of the after-shocks originated, but of which no trace appeared at the surface. The main fault of the Formosa earthquake of 1906 ended towards the west in a branch fault (Fig. 40). The Owens Valley earthquake of 1872 was connected with a system of faults. For one or two miles individual scarps maintain a nearly constant direction, but in some parts they are subject to abrupt changes of direction, so that their course is a succession of zigzags with sharp elbows; in others, they are arranged in parallel lines, slightly overlapping*.

79. Relation to the Form and Structure of the Ground. When viewed in detail, the course of an earthquake-fault seems to

* C. Davison, *Beitr. zur Geoph.*, vol. 12, 1912, p. 10; Omori, p. 57; Hobbs, pp. 374–383.

be independent of the form of the ground; and, in this respect, it differs totally from the earth-fissures described in Chapter VII. After running for considerable distances along a valley, the Mino-Owari fault crosses hill-spurs and in one case the top of a hill. Throughout its entire length, the San Andreas fault in California lies along depressions or at the base of steep slopes, which are due partly to erosion and partly to displacement along the fault. But its position with regard to the mountain-ridge frequently varies, as it passes several times through breaks in the chains from one flank to the other.

On a large scale, however, the course is dependent on the structure of the district. The Baluchistan earthquake-fault, for instance, runs in a nearly straight line parallel to the Khojak Range. The Locris earthquake-fault follows a nearly constant east-south-east direction parallel to the Gulf of Euboea. The Wellington earthquake-fault is represented by a continuous escarpment running along the foot of the Remutaka Mountains, where they present a steep slope towards the great Tertiary plain of the Wairarapa. One of the Alaskan earthquake-faults runs along the main portion of Russell Fiord (Fig. 39), another at the foot of the straight mountain front on the east side of Yakutat Bay*.

80. Horizontal Displacement. The horizontal displacement is usually shown by the dislocation of fences, roads, bridges, tunnels, pipes or any structure that crosses the line of the fault. Two examples of this displacement are illustrated in Figs. 34 and 37. In the former, the upper part of the severed road is shifted 13 feet to the left by the Mino-Owari fault; in the latter, the two portions of the fence which lie on opposite sides of the San Andreas fault are now separated by several feet.

Without a duplicated trigonometrical survey, it is usually impossible to determine which side has moved or whether both sides have moved. The relative displacement of the two sides is, however, so far as known, constant in direction throughout a given fault. In the Locris earthquake, the north-east side was shifted to the north-west relatively to the other side; and this was also the

* Koto, pp. 333, 336, 338, 341, 344; Lawson, vol. 1, pp. 48–52; C. Davison, *Geol. Mag.*, 1893, p. 359; S. A. Papavasiliou, *Compt. Rend. Acad. Sci. Paris*, vol. 119, 1894, p. 380; Lyell, p. 86.

case in the Mino-Owari earthquake. In the Baluchistan earthquake, the east side moved towards the north. In the Formosa earthquake, the north side was shifted relatively eastwards. Throughout the whole vast extent of the San Andreas fault, the displacement of the north-east side in 1906 was to the south-east.

The amounts of the horizontal displacement are sometimes considerable. In the Baluchistan earthquake, it was not less than $2\frac{1}{4}$ feet; in the Formosa earthquake it varied from 2 to 8 feet; in the Mino-Owari earthquake from 3 to 13 feet; in the

Fig. 37. Fence severed by the San Andreas fault-displacement in the Californian earthquake of 1906.

Owens Valley earthquake from 3 to 12 feet, and in places to 18 and 20 feet; in the Californian earthquake it lay as a rule between 8 and 15 feet, but in one place amounted to 21 feet.

In a few cases, the shifting parallel to the fault was accompanied by actual compression of the ground in the perpendicular direction. There is some slight evidence of this in the Baluchistan earthquake, and distinct proof of it in the Mino-Owari earthquake. Plots of ground that were 48 feet in length before the earthquake were reduced to 30 feet afterwards. Indeed, according to Milne, it would seem that the whole Neo Valley had become narrower*.

81. Vertical Displacement. The displacement of the ground in a vertical direction is usually conspicuous in the form of fault-scarps (Figs. 33, 34). Here, again, the measured displacement as a rule is relative only. A revision of the levels in a direction across the fault might determine which side, or whether both sides, had moved. In default of such precise evidence, the effects of the movement in changing the gradient of streams may be of service, as in the Mino-Owari and Assam earthquakes. In one case only, that of the Alaskan earthquakes of 1899, are the absolute changes of elevation known. The epicentral district being intersected by Yakutat Bay and its branches, the uplift could be measured even six years after the earthquakes by the heights of dead barnacles and mussels still clinging to the cliffs, and the subsidence by that of the base of the lowest dead tree in place.

The measured vertical displacements, whether relative or absolute, are in some cases less, in others far greater, than the horizontal movements. In the Californian earthquake the greatest uplift was 3 feet, in the Formosa earthquake 6 feet, and in the Wellington earthquake about 9 feet. In the Mino-Owari earthquake, the greatest uplift was about 10 feet, except at Midori (Fig. 34) where it amounted to nearly 20 feet. In the Owens Valley earthquake, one uplift of 23 feet was measured. The Chedrang fault-scarp of the Assam earthquake attained heights of 32 and 35 feet. The greatest known uplift is that of 47 ft. 4 ins. on the north-west shore of Yakutat Bay in Alaska.

In any one fault-scarp, the uplift may vary greatly. For instance, the length of the Chedrang fault mapped in Fig. 38 is about 12 miles, but, even in this short distance, there are three undulations, separated by intervals of no displacement, in which the maximum uplifts are 25, 35 and 32 feet, respectively. In a few cases, there may even be reversal in the direction of throw, as in the Mino-Owari fault-scarp at Midori (Fig. 34), where the north-east side was elevated nearly 20 feet, whereas, in all other

* C. Davison, *Geol. Mag.*, 1893, p. 359; Omori, pp. 57–59; Koto, pp. 330–346; Hobbs, pp. 379–380; Lawson, vol. 1, pp. 52–80; J. Milne, *Seis. Journ.*, vol. 1, 1893, p. 131.

parts of the known scarp, the uplift occurred on the south-west side*.

82. Surface Effects of Faulting. In the neighbourhood of fault-scarps, the ground is intensely fissured and landslips are almost continuous. But the most marked effects of faulting (and, to a less extent, of warping) are those on the flow of streams. If the uplift take place across a river-bed, the result is a decrease of gradient for a short distance upstream and an increase similarly downstream. The former effect is the more important, the river being often widened out or ponded back by the obstruction into a small lake or pool. Such effects are common to all cases of faulting, but were unusually conspicuous along the course of the Chedrang fault formed during the Assam earthquake of 1897†.

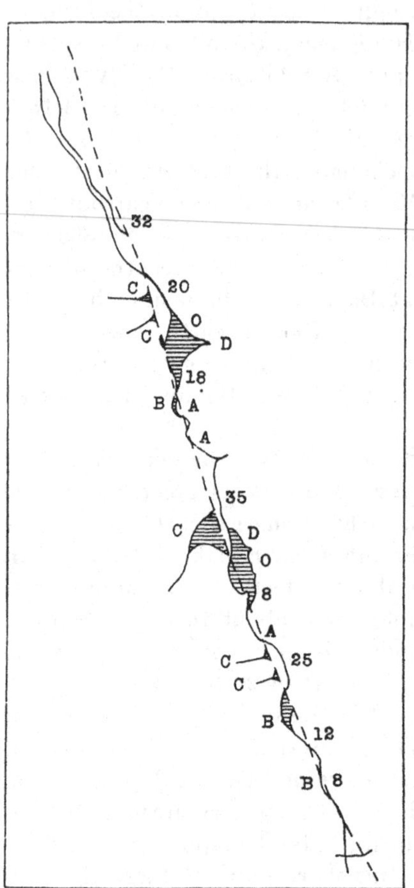

Fig. 38. Map of the Chedrang fault.

The course of this fault, as will be seen by the broken-line in Fig. 38, is practically straight for a distance of at least 12 miles. The figures on the right-hand side indicate the throw in feet in different parts of its course,

* Lawson, vol. 1, pp. 80–87, 140–145; Omori, pp. 57–59; Lyell, pp. 85–86; Koto, pp. 330–346; Hobbs, pp. 378–379; Oldham, pp. 138–148; Tarr and Martin, pp. 18–45.

† A photograph of one of these pools is reproduced in Marr's *Scientific Study of Scenery*, 1900, plate facing p. 180.

the rock on the east side, except where there is no displacement, being invariably the higher. Throughout its known course, the fault runs along the valley of the Chedrang river, which flows from south to north. Thus, whenever the stream crosses the fault-line from east to west there is a waterfall, as at *A*, etc. When the stream meets the fault-scarp from the west, it is ponded back and forms pools, as at *B*, etc. Pools also collect on the west side of the fault when small tributary streams meet the scarp, as at *C*, etc.

Besides these pools, there are others, at *D*, *D*, with a different origin. Each is about half a mile long, there is no visible barrier, and in one case the pool spreads across the fault. In both, the channel of the stream sinks in the upstream direction, without trace of faulting, beneath the waters of the pool. Now, it is just where the fault has no throw that the pools are widest and deepest. The pools must therefore be due to warping or the formation of an undulation across the stream-course, by which the natural slope of the ground has been reversed*.

Nature of Fault-Deformation

83. The nature of the fault-deformation is definitely known in at least eleven earthquakes, but, even in this small number, it varies considerably. There seem to be at least four kinds of deformation. (i) In the first class, the movement is almost entirely horizontal, or the horizontal displacement is greatly in excess of the vertical. To this class belong the Californian, Sumatran, Baluchistan, and possibly, the Locris, earthquakes. (ii) In the second class, the vertical displacement predominates or occurs without any horizontal shifting. The Wellington, Assam, Alaskan, and possibly the Owens Valley, earthquakes are typical of this class. (iii) The third class includes the Mino-Owari and, possibly, the Sonora, earthquakes, in which both horizontal and vertical movements occur to approximately the same extent. (iv) The fourth class includes, so far as is known, only the Formosa earthquake and is one of great interest, the vertical displacement being in opposite directions in the two halves of the fault.

* Oldham, pp. 138–148.

84. Movements chiefly Horizontal. In the Californian earthquake, the vertical displacement was small compared with that in a horizontal direction, and to the south of San Francisco was imperceptible. No vertical displacement was detected with the Sumatra earthquake. In the Baluchistan earthquake, the only measurement of uplift amounted to about 2 inches.

The question as to which side moved or whether both sides moved is one that can only be answered by a renewal of the trigonometrical survey of the district, and this was accomplished for the Californian earthquake before the lapse of a year. The area covered by the survey is about 170 miles long and 50 miles in maximum width. It extends from a line a short distance to the south of San Juan to the neighbourhood of Fort Ross (Fig. 36). It was found that since 1851, when the first survey was begun, the stations in the neighbourhood of the fault had all been displaced by amounts ranging from less than a foot to 19½ feet.

Though more than half a century had elapsed between the dates of the two surveys, there can be little doubt that the movements revealed by the re-triangulation occurred in 1906. (i) The horizontal displacements at the various stations were nearly always in directions parallel to the fault. (ii) The total displacements of stations close to the fault were approximately equal to those observed in lines of road or fencing severed by the fault.

The most important points established by the new survey are: (i) that both sides of the fault were displaced, the south-west side to the north-west, and the north-east side to the south-east; and (ii) that the movements were not confined to the immediate neighbourhood of the fault, the displacement decreasing on both sides with increasing distance from the fault.

For instance, on the north-east side of the fault, ten points at an average distance of a little less than a mile from the fault have an average displacement of 5·1 feet to the south-east; three points at an average distance of 2½ miles have moved on an average 2·8 feet in the same direction; while one point at a distance of 4 miles has a displacement of 1·9 feet. No point on this side at a greater distance than 4 miles has suffered any displacement distinctly exceeding that due to errors of observation. On the south-west side of the fault, twelve points at an average distance of 1¼ miles have an average displacement of 9·7 feet to the north-west; seven at an average distance of 3½ miles have one of 7·8 feet; while one point 23 miles from the fault was displaced 5·8 feet. Thus, straight lines drawn on the surface on either side of the fault and at right angles to it would after the earthquake become slightly curved, the concavity facing the south on the north-east side, and facing the north on the south-west side. Moreover, for points on opposite sides of the fault and at equal distances from it, the displacements on the south-west side were twice as great as those on the north-east side up to a few miles from the fault.

The first earthquake in which sudden displacements were afterwards established by geodetic measurements was the Sumatra earthquake of 1892. The movements were then entirely horizontal. There was no trace of any visible fault at the surface, but the measured displacements, according to H. F. Reid, show that the fault must be directed to the north-north-west, that the crust on the west side was shifted to the north and that on the east side towards the south; that the total relative displacement of the two sides was 11½ or 13 feet; and that the displacement diminished rapidly with increasing distance from the fault*.

The results of this section may thus be summed up as follows. So far as at present known, (i) the movement when chiefly horizontal is confined to strike-faults; (ii) both sides move in opposite directions; (iii) the amount of the displacement diminishes rapidly with increasing distance from the fault.

85. Movements chiefly Vertical. In the Assam earthquake, the movements along the two principal faults were entirely vertical. In the Alaskan earthquake, the vertical movements predominated, though there may in parts have been small horizontal displacements. An uplift only is recorded during the Wellington earthquake of 1855; but, in this early case, the evidence does not preclude the occurrence of small horizontal shifts. Horizontal movements certainly took place during the Owens Valley earthquake, but there is no evidence to show that they were more than local.

Of these four earthquakes, the Wellington earthquake was

* Lawson, vol. 1, pp. 114–145; Reid, pp. 72–78.

marked by the simplest type of displacement. On the west coast of North Island, there was no perceptible uplift at a point 16 miles to the north of Wellington. To the south of this point and eastwards along Cook Strait, the amount of upheaval increased gradually until it reached 9 feet on the east side of Port Nicholson and along the east flank of the Remutaka Mountains, where it formed a fault-scarp 90 miles in length. At the

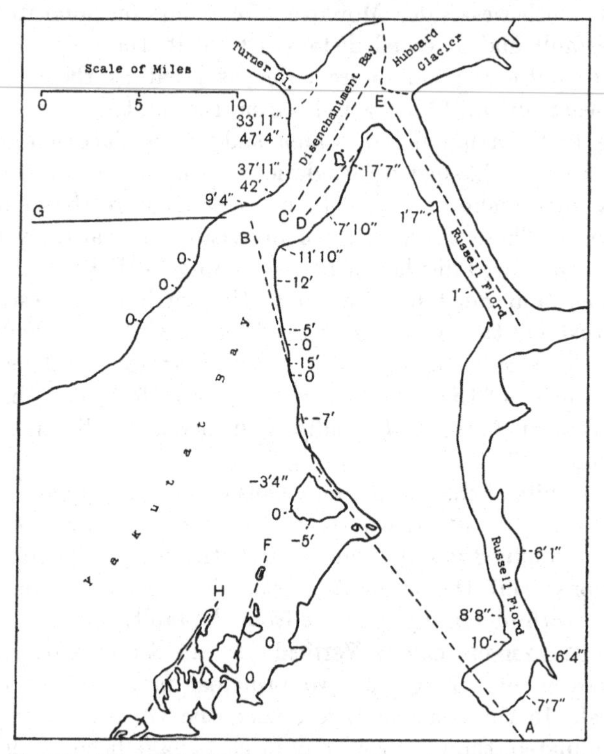

Fig. 39. Map of the Yakutat Bay earthquake-faults in 1899.

southern end of this scarp, the uplift of 9 feet was measured with reference to the level of the sea by the elevation of a white band of nullipores; and the same evidence showed that, to the south at any rate, the Wairarapa plain on the east side of the scarp was undisturbed. The effects of the movements were thus to uplift the Remutaka Mountains by about 9 feet, and to tilt a tract of country, 90 miles long and 23 miles wide, slightly to the westward.

The displacements of the Alaskan earthquake were of a more complex character. Along the coast, they were probably confined to Yakutat Bay and its branches (Fig. 39). The total length of coast-line of this inlet is about 150 miles. For a third of this distance, there was either no change or a very small change of level. These portions are indicated in Fig. 39 by small circles. Elevated regions, by far the more numerous, are denoted by figures only, the depressed regions by figures with minus signs. The figures show the amount of the uplift or subsidence in feet and inches.

It will be noticed how rapidly the amount of the uplift varies within a short length of coast. For instance, on the west coast of Disenchantment Bay, the uplift at one point is 42 feet, about a mile to the west 30 feet, and a quarter of a mile farther on only 9 feet. Though no great fault-scarps were noticed, there can be little doubt that these rapid variations were due to faulting rather than to warping, in part perhaps to a series of minor faults like that shown in Fig. 33.

The broken-lines in Fig. 39 represent the probable courses of the faults which, according to Tarr and Martin, are implied by the variations in the changes of level. These faults divide the crust into at least three distinct blocks, the known sides of which are roughly parallel. One of these blocks is bounded on three sides by the faults, *A*, *B*, *C* and *E*, a second by the faults *C* and *G*, while the third is bounded by the fault *E* and includes the north-east shore of the main portion of Russell Fiord. All three blocks extend to unknown distances in other directions than those mentioned. The principal effects of the widespread movements were an uplift of all the blocks along the lines of fault and a slight tilting of the masses away from the faults. The uplift was, however, accompanied by other movements—by a slight depression on the west side of the fault or faults *A*, *B*, and in many places by small slips along series of minor faults, due apparently to local adjustments in the tilted blocks*.

The results of this section may be summarised as follows: (i) the movements when chiefly vertical are confined to vertical faults or "blatts"; (ii) either both sides move in opposite directions, or one (the mountainous) side alone is moved and

* Lyell, pp. 82–89; Tarr and Martin, pp. 18–45.

uplifted; (iii) the effect of the uplift is to tilt one or more large crust-blocks in the direction away from the fault or faults.

86. Vertical and Horizontal Movements. The Mino-Owari earthquake was associated with a transverse fault which crossed almost the whole width of the Main Island in a general north-west and south-east direction. The relative displacement of the north-east side of the fault in the horizontal direction was invariably to the north-west and usually ranged from 3 to 13 feet, and in the vertical direction downwards by amounts varying from 0 to 10 feet. In the neighbourhood of Midori, however, or near the centre of the fault, the throw of the fault was reversed, the north-east side being raised about 20 feet above the other*.

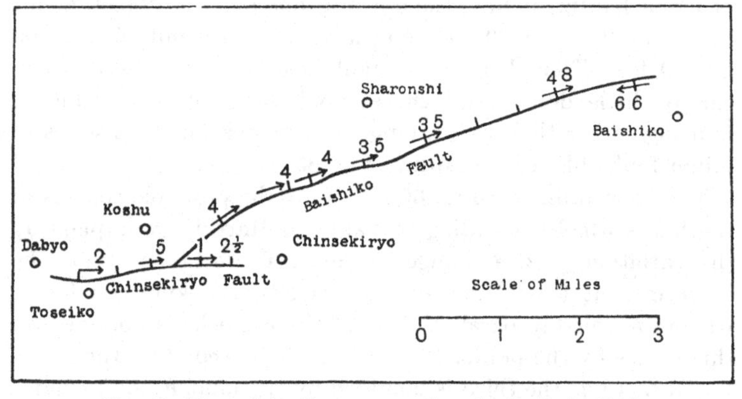

Fig. 40. Map of the Formosa earthquake-faults in 1906.

87. Complex Displacements. The only evidence with regard to such displacements is that furnished by the Formosa earthquake of 1906. The plan of the fault, so far as traced, is shown in Fig. 40. The length of the main portion is about 7 miles and of the branch about 2½ miles; but Omori thinks it probable that the branch-fault extended about 7½ miles farther west, and that the form of the isoseismal lines indicates an extension of the main fault at least 12 or 16 miles farther east in a mountainous tract, so that the total length of the fault would be about 30 miles. Throughout the whole observed length of the fault, the north side was displaced relatively to the east by amounts ranging from 2 to 8 feet; and, throughout the greater

* Koto, pp. 330–346.

part (indicated by the short perpendicular lines), the north side was depressed relatively by 3 or 4 feet. Towards the east end, however, the south side was depressed with reference to the other by as much as 6 feet. Assuming this condition to be continued throughout the inferred continuation of the fault, the complex displacement may be represented by Fig. 41, in which the shaded areas denote the portions depressed with reference to the adjoining portions on the other side of the fault, while the arrows indicate the directions of relative shear. The significance of this form of displacement will be seen when we consider the phenomena of twin earthquakes (sects. 242–244)*.

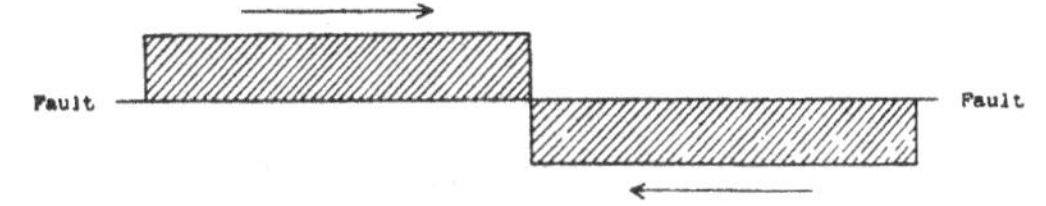

Fig. 41. Diagram illustrating the displacements in the Formosa earthquake of 1906.

Warping

88. In warping, the variation in the amount of displacement is continuous; there is no sudden break along a definite line, as there is in faulting. The warping may be (i) general, or extending over a considerable area, as in the Kangra earthquake of 1905 and the Messina earthquake of 1908, or (ii) local, as in the New Madrid earthquakes of 1811–1812 and (in part) in the Assam earthquake of 1897.

89. General Warping. The Kangra earthquake of 1905 originated in two foci (Fig. 52), one beneath Kangra and Dharmsala, the other beneath Dehra Dun and Mussoorie, the intensity of the shock at the former towns being far greater than at the latter. There was, however, no visible sign of disturbance at the surface, and, from this and from the uniform distribution of damage in the central areas, it may be inferred (sect. 139) that both the foci were deep-seated. The only line of levels made before the earthquake was that of 1862 from Saharanpur (on the plain) through Dehra Dun to Mussoorie. The portion from Dehra Dun to Mussoorie was repeated in 1904, less than a year before the earthquake, and again about a month after it, and

* Omori, pp. 57–59, 70–72.

these observations indicated that, in the interval, the height of Dehra Dun with respect to Mussoorie had increased by 5 inches. A fresh series of levels, carried out along the whole line from Saharanpur in 1906–1907 showed that, regarding the height of Saharanpur as fixed, Dehra Dun had risen about 5 inches, while the height of Mussoorie was almost unchanged. Though the amount is a small one, it seems reasonable to conclude that the only superficial effect of the deep-seated movement within the Dehra Dun focus was this very slight buckling-up of the crust.

The Messina earthquake of 1908 had also two foci, both lying beneath the Straits of Messina. A line of levellings had been carried round the Calabrian coast during the years 1906–1908, and this was repeated about two or three months after the earthquake from Giaio Tauro, which lies about 20 miles north-east of San Giovanni at the northern entrance to the straits, to Porto Salvo on the south coast. Assuming the height of the former place to have been unaltered, the measurements indicate a subsidence of the whole east coast of the straits, of 17 inches at San Giovanni, 21 inches at Reggio, 24 inches 2 miles south of Reggio, and 14 inches at Pellaro at the southern entrance to the straits. On the opposite coast, the subsidence was slightly greater, amounting to 28 inches at Messina*.

90. Local Warping. During the New Madrid earthquakes of 1811–1812, local warping took place on an extensive scale. According to Fuller, the earthquakes were caused by displacements along a fault lying about 15 miles west of the river Mississippi, directed about N. 30° E., and about 75 miles in length. In the central district of the earthquake, there are three (if not four) linear troughs each still occupied, though not continuously, by "sunklands," consisting of river-swamps and lakes. The greatest depths of the hollows in which these lakes are situated are from 15 to 20 feet. The troughs are separated by two (if not three) discontinuous ridges, trending in nearly the same direction as the inferred fault. On one of these ridges lie three low flat domes, the largest of which is about 15 miles long and from 5 to 8 miles wide, the surface having been uplifted from 15 to 20 feet. The total length of this ridge is not less than 70 miles.

* C. S. Middlemiss, *Mem. Geol. Surv. India*, vol. 38, 1910, pp. 348–349; *Rel. della Com. Reale*, pp. 131–156.

It should be noticed that, in this case, the observed warping is that of the alluvial beds of the Mississippi plain, and not of the solid crust below. But warping on so large a scale and in so uniform a direction can hardly exist in the surface layer alone. It implies deformation of the crust below, though whether that deformation be in the form of warping or faulting or both is of course unknown.

Some interesting examples of local warping occurred in connexion with the Assam earthquake of 1897. At several places within its central area, the form of the hills was perceptibly changed. In other parts, surface undulations altered the gradient of streams, and gave rise to pools or small lakes. Two such pools along the Chedrang fault have been described in sect. 82 (Fig. 38, *D*, *D*). Other groups of pools were formed at a distance from any visible fault, some of them from a quarter of a mile to a mile in length, and one of them more than 20 feet deep.

The epicentral area of the Assam earthquake was of great magnitude—according to Oldham, about 200 miles long from east to west and not less than 50 miles in width (Fig. 47). Near the boundary, the displacements were probably long low rolls, resulting in changes in the aspect of the hills; these were succeeded by more pronounced undulations sufficient to reverse the drainage of rapidly-flowing streams; and these again merged in the central regions into fractures and faults*.

91. Probable Connexion between Warping and Faulting. The distribution of surface warping and faulting in the Assam earthquake suggests that there may be some connexion between the two phenomena. Fractures, which at the surface show no displacement, may have a considerable throw a mile or two below it; and the undulations that gave rise to pools are probably mere folds in the upper rocks produced by faulting below. The slope illustrated in Fig. 35, for instance, is merely an undulation in a thick bed of alluvium due to the passage of the Chedrang fault below. Thus, when faulting takes place at some depth, it is represented either by no displacement or by warping of the outer crust. It is only when the fault-displacement occurs at a small depth that scarps are left projecting at the surface.

* Fuller, pp. 15, 62–75, 105–109; Oldham, pp. 138–163.

CHAPTER VI

SEISMIC SEA-WAVES

92. Under the general term seismic sea-waves (*tsunamis* of the Japanese seismologists) are included waves of three kinds: (i) Condensational waves, usually known as seaquakes; (ii) Gravitational waves, which, under certain conditions, are propagated in all directions from a submarine epicentre with a velocity depending on the depth of the sea; (iii) Stationary waves, the periods of which in different bays are governed by the forms and dimensions of the bays. These are also called *marine seiches**.

Of these three forms of sea-wave, the first occur with many earthquakes, whether their origin be inland or submarine, and in lakes and rivers as well as in the sea; the second only with great earthquakes of submarine origin; while the third may be excited in bays by the preceding class of waves, but frequently occur at other times from the action of storms, changes of barometric pressure, etc.

The first class of waves or seaquakes consist of condensational vibrations only, water being incapable of transmitting distortional vibrations. In earthquakes of moderate strength, the surface of the water in pools or the sea is ruffled. In strong

* The principal memoirs on seismic sea-waves are the following:

1. Geinitz, F. E. Das Erdbeben von Iquique am 9. Mai 1877, etc. *Ksl. Leop.-Carol.-Deutschen Akad. der Naturf.* (Halle), 1878, pp. 385–444.

2. Hochstetter, F. von. Über das Erdbeben in Peru am 13. August 1868, etc. *Sitzungsb. Akad. Wien*, vol. 58, 1868, pp. 837–860; vol. 59, 1869, pp. 109–132; vol. 60, 1870, pp. 818–823.

3. Honda, K., Terada, T., Yoshida, Y., and Isitani, D. Secondary undulations of oceanic tides. *Publ. Eq. Inv. Com.*, No. 26, 1908, pp. 1–113.

4. Milne, J. The Peruvian earthquake of May 9th, 1877. *Trans. Seis. Soc. Japan*, vol. 2, 1880, pp. 50–96.

5. Platania, G. Il maremotto dello Stretto di Messina del 28 dicembre 1908. *Boll. Soc. Sis. Ital.*, vol. 13, 1908, pp. 369–458.

6. Wharton, W. J. L. On the seismic sea-waves caused by the eruption of Krakatoa, August 26th and 27th, 1883. *The Eruption of Krakatoa and Subsequent Phenomena* (edited by G. J. Symons). 1888, pp. 89–150.

earthquakes, the effect on ships is as if they had struck on a rock or grated over a reef, and the surface of the water is strongly or tumultuously disturbed. The principal value of sea-quakes consists in the light which they throw on the distribution of submarine earthquakes (sect. 184). In the present chapter, waves of the second and third classes of seismic origin will alone be considered. They will be distinguished here as seismic sea-waves and seismic seiches.

93. Frequency of Seismic Sea-Waves. The number of earthquakes accompanied or followed by seismic sea-waves is apparently small. In Mallet's catalogue of recorded earthquakes*, 47 earthquakes during nearly three and a half centuries (1501–1842) are described as accompanied by sea-waves. In about twelve centuries (684–1897), 35 sea-waves were recorded in Japan in connexion with earthquakes. In all these cases, however, the sea-waves were of considerable magnitude. The total number must be far larger. Though it is at present impossible to give exact figures, it seems probable that the proportion of submarine earthquakes accompanied by sea-waves does not differ greatly from the proportion of inland earthquakes that are accompanied by the formation of fault-scarps.

94. Nature of Seismic Sea-Waves at Places near the Epicentre. Soon after the earthquake, at intervals varying from a few minutes to half an hour or more, a great sea-wave approaches the land in one long unbroken ridge. As it passes into shallow water, the front of the wave becomes steep, then impending, and finally it breaks and sweeps over the low-lying ground. The general testimony of observers is that the first movement on the shore is a retreat of the water, which lasts for 5 or 10 minutes or even longer before the arrival of the waves. In some cases, as in the Riviera earthquake of 1887, this observation is confirmed by the records of neighbouring tide-gauges.

With regard to the height of the waves, personal testimony is usually untrustworthy. At the same place, estimates may vary from 20 to 80 feet, and it is uncertain whether they refer to the height of the crest above the previous level of the sea or above the trough of the succeeding wave, that is, to the amplitude or the range of the waves. There can be no doubt, however, that

* *Rep. Brit. Ass.*, 1852, pp. 1–176; 1853, pp. 118–212; 1854, pp. 1–326.

the height is often considerable, and estimates can hardly be regarded as excessive which attribute a height of 30 feet to the sea-waves of the Alaskan earthquake of 1899, or a height of 50 or 60 feet to those of the Iquique earthquake of 1877 or the Lisbon earthquake of 1755, or 93 feet to the Sanriku (Japan) earthquake of 1896. In the Messina earthquake of 1908, which was far less intense than any of these earthquakes, Platania measured the heights of the waves from remains left on the walls or on the ground, and found that the maximum height on the Sicilian shore was 28 feet, and, on the opposite Calabrian coast, 35 feet between Pellaro and Lazzaro*.

The height of the waves, however, does not depend entirely on the violence of the shock. It is governed partly by the inclination of the sea-bed in the neighbourhood of the shore, partly by the form of the coast, being naturally great at the head of a gradually narrowing bay. In the Messina earthquake of 1908, the greatest heights were attained at some distance from the epicentral districts. In the Valparaiso earthquake of 1906, no sea-waves were noticed at Valparaiso itself owing to the great depth of water along the coast, while, at Kushimoto in Japan (10,937 miles from Valparaiso) the total range of the movement was 4 inches.

The first sea-wave is usually followed by several others, as a rule, but not always, of less magnitude. After the Concepcion earthquake of 1835, three waves swept over the town; in the Japanese earthquake of 1854, there were seven or eight waves; in the Iquique earthquake of 1877, the sea broke eight times over both Iquique and Arica; the Lisbon earthquake of 1755 was followed by 18 waves at Cadiz.

One of the most carefully observed series of sea-waves were those associated with the Sanriku (Japan) earthquake of June 15, 1896. The shock was felt at Miyako at 7.32 p.m. About 7.50, the sea began to retire; it then rose until about 8.0 and again retired. At 8.7, the largest wave invaded Miyako, and other large waves swept the shore at 8.15, 8.32, 8.48, 8.59, 9.16 and 9.50 p.m. The intervals between successive waves were thus 7, 8, 17, 16, 11, 17 and 34 minutes; from which it may be concluded that the waves had periods of about 16 and 8 minutes.

* Platania, pp. 371–428.

The period of the oscillations commonly observed in the Bay of Miyako in ordinary weather is somewhat different, namely, 21 minutes*.

95. Effects of Seismic Sea-Waves. A few examples of the power of seismic sea-waves may be given. The waves referred to in the last paragraph destroyed the town of Kamaishi and swept many of the houses into the sea. Two schooners were left among the ruins, one of them 200 yards from the shore. On the east side of the Bay of Yakutat in Alaska, mature trees were wrecked by the sea-waves of 1899, the beach presenting a wild, almost impenetrable, tangle of uprooted, broken, twisted and shattered trunks. The Messina sea-waves of 1908 wrought most damage on the Calabrian coast of the Straits. At Reggio, a block of concrete, 8½ feet long, 8 feet wide and 4 feet thick, was torn from the pier and carried more than 20 yards by the waves. The sandy shore between Pellaro and Lazzaro was also swept away, in places to a width of more than a hundred yards†.

96. Distances traversed by Seismic Sea-Waves. The earthquakes on both sides of the Pacific Ocean generate sea-waves that are observed on the opposite coasts. The waves of the Japanese earthquake of 1854 were recorded at San Francisco (5089 miles) and San Diego (5593 miles); those of the Sanriku earthquake of 1896 at the former place (4787 miles). Much greater are the distances traversed by the waves due to South American earthquakes. The Peruvian earthquakes of 1868 and 1877 were registered by tide-gauges at Hakodate in Japan (10,315 miles), those of the Ecuador earthquake of 1906 at Fukahori in Japan (9992 miles), and those of the Valparaiso earthquake of 1906 were recorded at Kushimoto in Japan (10,937 miles). The Krakatoa sea-waves of 1883, though not of seismic origin, were recorded at Havre (12,867 miles). The great distances traversed by sea-waves are of course due to the fact that the waves diverge practically in two dimensions only.

97. Nature of Seismic Sea-Waves at great Distances. As the sea-waves diverge from the epicentre, they diminish rapidly in amplitude, becoming long low swells, 100 or 200 miles in length.

* Honda, pp. 90–91.

† R. S. Tarr and L. Martin, *U. S. Geol. Surv.*, Prof. Paper No. 69, 1912, pp. 46–48; F. Omori, *Bull. Eq. Inv. Com.*, vol. 3, 1909, pp. 41–42.

At distances of a few thousand miles, they no longer retain the appearance of sea-waves, and approximate to that of tidal waves of short period, from 20 to 30 minutes in duration, the sea slowly rising and sweeping up the beach and then as slowly falling. It is no doubt from this gradual rise and fall that seismic sea-waves have been erroneously called tidal waves. For instance, in Hakodate Bay (Japan), the Iquique sea-waves of

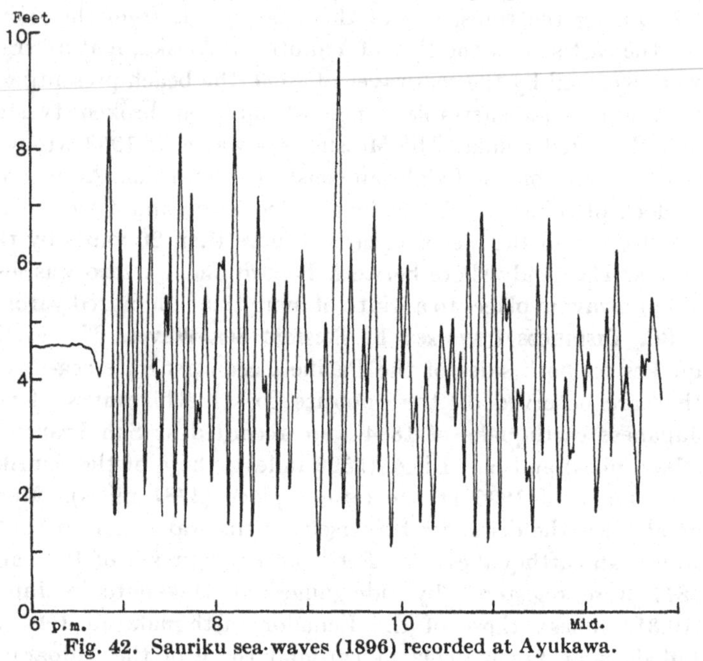

Fig. 42. Sanriku sea-waves (1896) recorded at Ayukawa.

1877, after travelling 10,315 miles, appeared as miniature tides with a period of about 20 minutes, a maximum amplitude of nearly 8 feet, and a total duration of several hours*.

98. Nature of Seismic Seiches. The nature of seismic seiches, as recorded by tide-gauges, will be evident from Figs. 42, 43. Fig. 42 is a reproduction of part of the record of the Sanriku waves of 1896 at Ayukawa, 168 miles from the epicentre. Fig. 43 represents the seiches due to the Iquique earthquake of 1877, as recorded at San Francisco, 5240 miles from the epicentre. The records differ of course chiefly in the amplitude of the

* Honda, pp. 84–86.

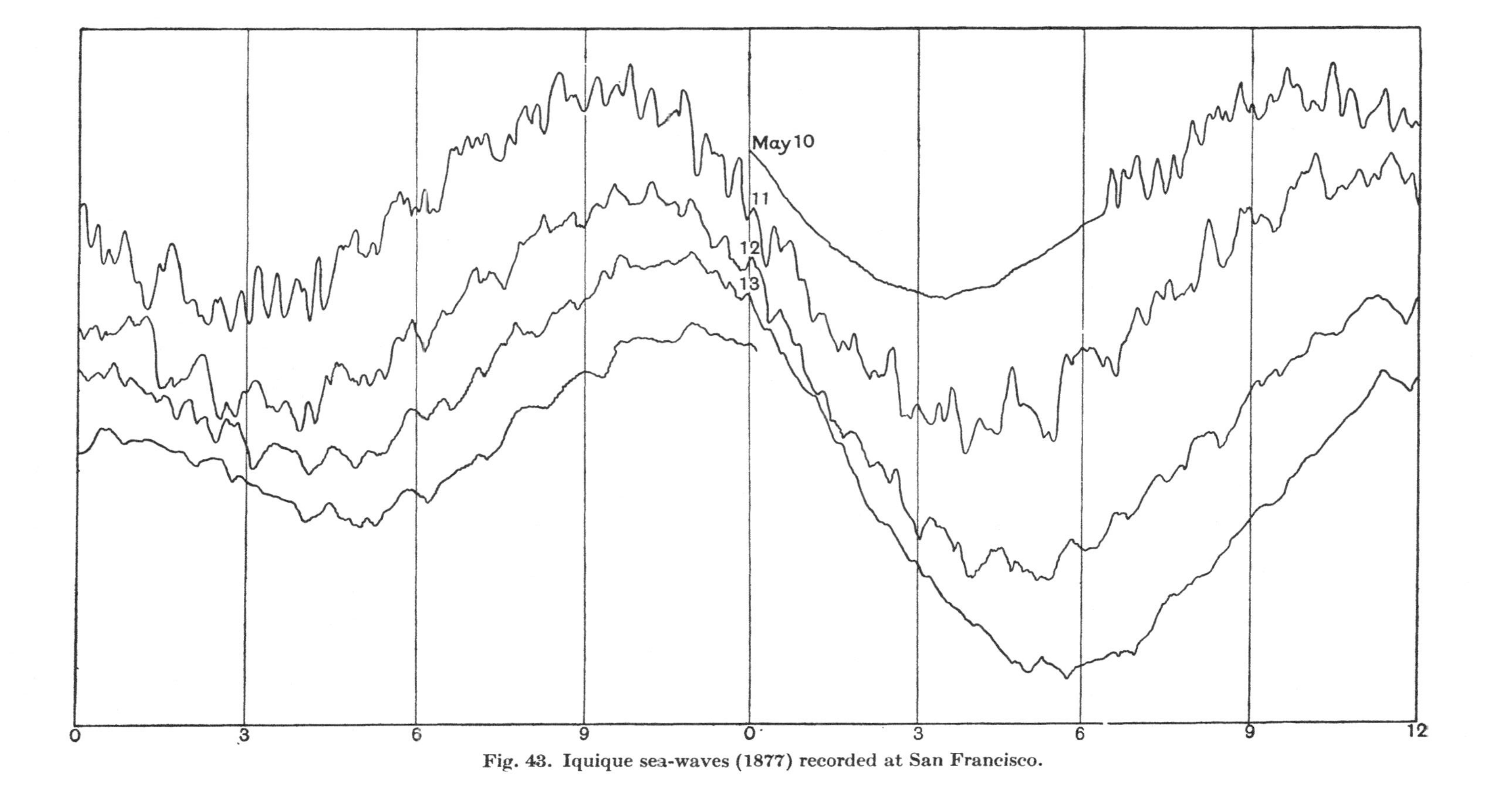

Fig. 43. Iquique sea-waves (1877) recorded at San Francisco.

oscillations and in their respective ratios to the amplitude of the tidal wave. The San Francisco record (Fig. 43) shows the principal features of seismic seiches: (i) their long duration, in this case for more than two days; (ii) the prevalence of certain periods in the oscillations, the average periods of comparatively regular series being 17·3, 27·8, 34·3 and 47·4 minutes, the most conspicuous being 34·3 minutes and the octave 17·3 minutes; and (iii) the relation of one of these periods (34·3 minutes) with that of one of the periods of oscillation (38–48 minutes) of the water of San Francisco Bay.

99. Periods of Seismic Seiches. The similarity of seismic sea-waves in bays to the seiches observed in lakes was pointed out by Omori in 1901, when he discovered that the waves recorded in each bay, whether excited by earthquakes or by storms, have their own proper period or periods. He inferred that each enclosed portion of the sea is virtually a fluid pendulum, the oscillations of which are fixed in period by the form and dimensions of the bay. Elaborate observations on marine seiches have since been made by K. Honda and his colleagues.

Some examples of the regular periodicity of seismic seiches may be given.

Hakodate Bay (Japan) is semicircular in form. The periods of the most conspicuous seiches vary from 45·5 to 57·5 minutes and from 21·9 to 24·5 minutes, the former corresponding to the fundamental oscillation of the bay, the latter to its lateral oscillation. The periods of the Sanriku seiches (1896) were 18·8, 39·5 and 57·5 minutes, and those of the smaller seiches due to an earthquake of 1897 were 22·1 and 45·5 minutes. The periods of the Ecuador seiches of 1906 were 21·9 and 40·9–49·2 minutes, and those of the Valparaiso seiches of the same year 22·1 and 48·0–53·0 minutes.

At Honolulu, the average periods of the Krakatoa seiches (1883) were 27·7 minutes, of the Sanriku seiches (1896) 23·4–26·0 minutes, of the Ecuador seiches (1906) 24·8–26·8 minutes, and of the Valparaiso seiches (1906) 26·2 minutes. All of these periods agree closely with that of the fundamental oscillation of the inlet leading to Honolulu.

The Messina seiches of 1908 were recorded by nine tide-gauges in the Mediterranean, and, with one exception, their periods corresponded closely with those of seiches due to meteorological causes*.

The cause of the stable periodicity of marine seiches in any bay has been explained by Omori and by Honda and his colleagues. Considering first, for simplicity, the case of a rectangular bay of length l and uniform depth h, if waves approximately of period $4l/\sqrt{gh}$ be propagated from the ocean directly up the bay, such waves will be reflected at the head of the bay, and, by the interference of the incident and reflected waves, a stationary wave will be formed with its loop at the head, and its node at the mouth, of the bay†. Now, if waves of different periods proceed from the ocean towards the shore, the wave with a period coinciding nearly with that of the oscillation of the bay water will excite the most energetic oscillations of that water. Thus, the bays on a coast-line may be compared with a series of resonators, each of which selects, and resounds to, the note of its proper period from a chaos of sounds outside.

Besides the fundamental or uninodal oscillation referred to, oscillations with two, three, or more, nodes are possible, the periods of such oscillations being respectively one-third, one-fifth, etc., of that of the fundamental oscillation. In some cases, also, the lateral oscillation of the bay is possible, as in San Francisco Bay, in which the frequently observed period of 34·3 to 41·2 minutes is due to lateral oscillations between the West Berkeley and Sausalito sides of the bay.

In bays of regular shape, the position of the nodal line is determinate. In bays of complicated form, however, several nodal lines are possible, to each of which there will correspond a special period of the seiches. In such a bay, the seiches may thus have several different periods; and, as in the case of the Iquique seiches at San Francisco (Fig. 43), the periods may at times undergo changes from one epoch to another‡.

* D. Kikuchi, *Publ. Eq. Inv. Com.*, No. 19, 1904, p. 26; Honda, pp. 19–20, 90–92; Platania, pp. 423–428, 442–446.

† The distance between successive nodes of the wave is thus $2l$, and the length of the complete wave is $4l$. The velocity of the sea-wave, according to Lagrange's formula, being $\sqrt{gh}$, it follows that the period of the oscillation is $4l/\sqrt{gh}$.

‡ Honda, pp. 14–16, 77.

100. Mean Velocity of Seismic Sea-Waves. The velocity of a sea-wave at any point is known to depend on the depth of the ocean. Estimates of the mean velocity are therefore of little value unless the paths traversed by the waves be free from interruption by islands or shoals. The following table is therefore confined to those earthquakes in which the estimates seem most trustworthy, and to one wave-path in each case. The velocity of the Krakatoa sea-waves across the Indian Ocean is added for the sake of comparison *.

Earthquake	Station	Distance miles	Time-interval hrs.	mins.	Mean velocity miles per hr.	Authority
Japan, 1854	San Diego	5593	12	37	443	Honda
Peru, 1868	Hakodate	10315	24	57	413	,,
Sanriku, 1896	San Francisco	4787	10	34	453	Davison
Ecuador, 1906	Fukahori	9992	21	5	427	Honda
Valparaiso, 1906	Kushimoto	10937	23	31	465	,,
Krakatoa, 1883	Aden	4393	11	51	371	Wharton

101. Connexion between Mean Velocity of Sea-Waves and Mean Depth of Ocean. In an ocean of uniform depth h feet, the velocity of the seismic sea-waves, by Lagrange's formula, is approximately $\sqrt{gh}$ feet per second. Thus, the depth of the ocean, if it were uniform, could be determined from the known velocity of seismic sea-waves. For an ocean of variable depth, the corresponding value of the mean velocity, according to Davison's formula, is $s \div \int_0^s \frac{ds}{\sqrt{gh}}$ feet per second, s being the distance in feet between the epicentre and recording tide-gauge. Davison shows that the mean depth given by Lagrange's formula must invariably be less than the true mean depth, possibly by as much as 20 per cent. of the latter.

The difficulty in practice is to determine the course followed by the sea-waves. In the case, however, of the Sanriku earthquake (1896), the great circle joining the epicentre to Sausalito (San Francisco Bay) is entirely free from islands and crosses the sub-oceanic contour-lines approximately at right angles.

* Honda, pp. 80, 83, 94; C. Davison, *Phil. Mag.*, vol. 50, 1900, pp. 583–584; Wharton, table 1.

The distance between the epicentre and Sausalito being 4787 miles and the time-interval between the occurrence of the earthquake and the arrival of the sea-waves at Sausalito Bay 10 h. 34 m., it follows that the mean velocity between the two places was 664 feet per second. Now, the mean depth along the great circle is more than 17,000 feet, while that given by Lagrange's formula with the above value of the velocity would be 13,778 feet, or about four-fifths of the measured value*.

102. Origin of Seismic Sea-Waves. Any sudden disturbance of the ocean-bed, either by crustal movements or by submarine landslips, must give rise to sea-waves of greater or less magnitude. In a great earthquake, both causes may be in action. As a rule, however, it is probable that the abrupt formation of a submarine fault-scarp is the more frequent cause of seismic sea-waves of great magnitude. The following characteristic features of such waves seem to support this conclusion:

(i) The epicentres of earthquakes associated with sea-waves are invariably either wholly or in part submarine. No earthquake, however violent, with its epicentre entirely on land has ever given rise to seismic sea-waves; though seismic seiches may be produced in distant lakes by earthquakes of the first order of magnitude†.

(ii) Seismic sea-waves are associated as a rule only with earthquakes of great intensity. In the 47 cases of sea-waves recorded by Mallet from 1501 to 1842, the intensity of the shock is known in 24, and in each case, on the nearest land, it attained the degree 10 of the Rossi-Forel scale. Again, in the Japanese Empire, there have, since the earliest times, been ten very great earthquakes, of which three originated beneath the land, and seven beneath the Pacific Ocean, each of the latter having been

* C. Davison, *Phil. Mag.*, vol. 43, 1897, pp. 33–36; vol. 50, 1900, pp. 579–584. Wharton and Milne have obtained somewhat similar results for the sea-waves connected with the eruption of Krakatoa in 1883 and the Iquique earthquake of 1877 (Wharton, pp. 89–150; Milne, pp. 50–96); and G. Platania for those of the Messina earthquake of 1908 and the Ferruzzano earthquake of 1907 (Platania, pp. 446–450; *Boll. Soc. Sis. Ital.*, vol. 16, 1912, pp. 166–174).

† Seismic seiches were observed in many lakes and pools in Great Britain and other countries a few minutes after the occurrence of the Lisbon earthquake of 1755, and were not dependent on the submarine origin of the earthquake.

accompanied by destructive sea-waves. Forty other strong, but less violent, earthquakes had epicentres in the Pacific Ocean, but only 16 of these were associated with sea-waves. Thus, of the 47 strong or violent earthquakes with epicentres beneath the Pacific Ocean, 23, or nearly 50 per cent., gave rise to sea-waves. The earthquakes with epicentres beneath the Sea of Japan are much less numerous and of less intensity (sect. 179). Of 17 such earthquakes, only two were followed by sea-waves.

(iii) In a few cases, earthquakes associated with sea-waves have been accompanied by changes of elevation of the adjoining land; for instance, the Concepcion earthquake of 1835, the Alaskan earthquake of 1899 and the Messina earthquake of 1908 (sects. 81, 89). On the other hand, the fault-rift of the Californian earthquake of 1906 was in part submarine, and in this earthquake no seismic sea-waves were observed, the displacement being mainly or almost entirely horizontal (sect. 84).

(iv) Lastly, in two earthquakes with epicentres in small part submarine, the sea-waves, though large in the immediate neighbourhood of the epicentres, were too small in volume to affect tide-gauges at a moderate distance. The Alaskan sea-waves in 1899 were evidently generated in Disenchantment Bay and Russell Fiord. At one point on the east side of Yakutat Bay, mature trees were thrown down by them (sect. 95); but, 4 or 5 miles farther south, trees remained standing, and no trace of the sea-waves could be seen on the records of the tide-gauges at the mouth of the Yukon River or at San Francisco. The sea-waves of the Messina earthquake of 1908 were 35 feet high in the Straits of Messina, but they affected no tide-gauge in the Mediterranean at a greater distance than 820 miles.

Thus, while submarine landslips are known to occur—for the fractures of many deep-sea cables are due to them (sect. 184)—there can be little doubt that the great sea-waves, those that are propagated across the ocean to distances of 5000 or 10,000 miles from the epicentre, are caused by the formation of submarine fault-scarps comparable in magnitude with those of the Wellington and Mino-Owari earthquakes*.

* D. Kikuchi, *Publ. Eq. Inv. Com.*, No. 19, 1904, pp. 24, 26; R. S. Tarr and L. Martin, *U. S. Geol. Surv.*, Prof. Paper No. 69, 1912, pp. 46–48.

CHAPTER VII

SECONDARY EFFECTS OF EARTHQUAKES

103. The effects of earthquakes on land and water belong to two main classes. In the first are those connected with the formation of fault-scarps and other distortions of the crust, whether on land or under sea. These have been described in some detail in Chapters V and VI. The present chapter includes the effects which result from the shock itself and which have no connexion or very little connexion with the crust-displacement in which the earthquake originated. These are generally known as *secondary effects*. They include such phenomena as landslips of various forms, dislocations of alluvium on level ground, changes in the underground water circulation, the extrusion of water and sand at the surface, and the formation of sand-craters and other outflows of sand and materials contained in the superficial layers*

LANDSLIPS

104. Under the general term *landslip* are included: (i) *avalanches*, usually of rock and earth, occasionally of ice and snow; and (ii) *earth-slumps*, in which a large portion of ground drops away from the slope of which it formed a part, leaving a horse-shoe-shaped scarp overlooking the sunken area. The formation of earth-slumps depends to some extent on the amount of water saturating the mass. When this amount is excessive, the earth-slump merges into the *earth-flow*.

* On the secondary effects of earthquakes, reference may be made to the following memoirs:

1. Dutton, C. E. The Charleston earthquake of August 31st, 1886. *9th Ann. Rep. U. S. Geol. Surv.*, 1889, pp. 280–295.
2. Fuller, M. L. The New Madrid earthquake. *Bull. U. S. Geol. Surv.*, No. 494, 1912.
3. Lawson, A. C. (editor). *The Californian Earthquake of April* 18, 1906, vol. 1, 1908, pp. 384–409.
4. Oldham, R. D. Report on the great earthquake of 12th June, 1897. *Mem. Geol. Surv. India*, vol. 29, 1899, pp. 10–26, 85–123.

All these forms of landslips may and do occur without the aid of earthquakes. When resulting from earthquakes, they differ from those produced under normal conditions chiefly in their greater magnitude and the wider range of country over which they are to be found.

105. Avalanches. Avalanches are formed in hilly countries with every great earthquake. In the provinces of Mino and Owari (Japan), numerous landslips were precipitated by the earthquake of 1891. Long lines of mountain formerly green with forest looked afterwards as if they had been painted yellowish-white, while the valleys below were choked with debris. The Assam earthquake of 1897 was remarkable for the great development of landslips. In parts, they were so numerous that, from a distance, there seemed to be more landslip than untouched hill-side. Sandstone hills, almost covered with forest, were cleared of wood from crest to foot, and the sandstone stood out in an apparently unbroken stretch of 20 miles in length. In one small valley, the steep hill-sides were stripped bare, and in the bottom of the valley was a piled-up heap of debris and broken trees, which raised the level of the stream-course by many feet. Along the bluffs which face the Mississippi, the landslides caused by the New Madrid earthquakes of 1811–12 are still conspicuous for a distance of at least 35 miles. "Sharp ridges of earth alternate with deep gashes..., the whole surface locally being broken into a jumble of irregular ridges, mounds and hummocks, interspersed with trench or basin-like hollows and other more irregular depressions." This is the usual aspect of the landslips, but, in some cases, the landslip descends with unbroken surface; as, during the Calabrian earthquakes of 1783, when large masses of clay fell from cliffs 500 feet in height and were deposited on the floor of the ravines with trees and crops still growing on them, and even with houses uninjured*.

106. Conditions governing the Formation of Landslips. The conditions that govern the widespread occurrence of landslips have been summarised by Oldham as follows: (i) the violence of the earthquake; (ii) the natural tendency of the hill-side to slip, which obviously varies with the slope of the surface;

* J. Milne, *Seis. Journ.*, vol. 1, 1893, p. 132; Oldham, pp. 114–120; Fuller, pp. 59–61; R. Mallet, *Rep. Brit. Ass.*, 1850, p. 49.

(iii) the height of the slope from crest to base, the greater swing being imparted to the higher hills, and (iv) the mineral constitution of the hill and the sharpness of the boundary between the weathered soilcap and the underlying rock. The importance of the third condition was manifested during the Assam earthquake of 1897, landslips being almost entirely absent from districts in which the hills are low, but common in adjoining parts in which the hills are high and the valleys deep*.

107. Effects of Landslips on Watercourses. (i) A common effect of landslips is the diversion of streams into new courses. This is increased by the material denuded from the exposed hill-sides. After the Assam earthquake, the burden of sand thus brought down into the streams was far greater than they could carry away. In consequence, the beds of the rivers were raised; and the rivers, instead of being torrents when in flood or a succession of pools and rapids at other times, became broad shallow streams flowing over thick deposits of sand.

(ii) Landslips from one or both sides of a valley frequently shoot across the river-bed and pond back the water; but lakes so formed are rarely permanent. One formed during the Mino-Owari earthquake of 1891 was 6 miles in circumference. One unusually large landslip of the Assam earthquake formed a barrier of which the remains are more than 200 feet above the present level of the river-bed. Above this barrier, the river accumulated in a great lake which lasted for about 3 months, after which the barrier burst and a great flood rushed down the valley†.

108. Earth-Slumps. During the Californian earthquake of 1906, earth-slumps were the most common form of landslip, and many attained a great size. The largest occurred to the south of Cape Fortunas. One month after the earthquake, it projected into the ocean for a quarter of a mile, forming a new cape, its length, measured in the direction of its flow, being nearly a mile, and its greatest width about half a mile. At its head, from which the mass broke away, were a number of steep scarps from 100 to 200 feet in height.

In the formation of earth-slumps, water plays a considerable

* Oldham, pp. 112–114.

† Oldham, pp. 120–121; J. Milne, *Seis. Journ.*, vol. 1, 1893, p. 132.

part, the masses, when saturated, being often on the point of motion. In many cases, the small residual stability of the earth-slump was overcome by the vibration of the ground. To a less extent, the movement was aided by the water expelled from the ground at the time of the earthquake, and also by the opening of fissures by means of which the rains of the following winter were able to reach the deeper portions of the mass.

When the sudden accession of water to the unconsolidated materials on a slope is considerable, earth-slumps merge into *earth-flows*, in which the plastic soil creeps like a lava-stream, and forms a fan or tongue of debris on the slope below. The movement in earth-flows is more rapid than in slumps, the whole process of briefer duration, and, unlike that of slumps, the action is final and non-recurring*.

Compression of Alluvium

109. Earth-Lurches. A fundamental feature of avalanches and earth-slumps is their limitation to more or less steeply-sloping ground. The displacements of alluvium described in the present section are practically confined to level ground. And it is only in earthquakes of great intensity that they are common or distinctly marked.

In the Californian earthquake of 1906, they were chiefly formed in the alluvial plains bordering on streams. In the simplest case, they were merely fissures accompanied by a slight movement of the ground on one or both sides. In their extreme form, the ground was cut up by fissures into strips or prisms which lurched towards the stream-trench, the lurching being usually accompanied by rotation.

After the Assam earthquake, telegraph-poles, originally set in a straight line and far away from river-channels, were found to be displaced as much as 10 or 15 feet. Along the Assam-Bengal railway, the line at one spot was shifted laterally 6 ft. 9 ins., the total length of line affected being nearly half a mile. In this case, the ground on either side of the line as well as the railway-embankment shared alike in the lateral shift, for there were no fissures at the foot of the bank.

A different form of earth-lurch is widely manifested in dis-

* Lawson, pp. 385–386, 390, 394.

tricts in which there are no stream-trenches to aid in its formation. During the Californian earthquake, the tidal mud-flats of Tomales Bay and the "made land" of San Francisco were thrust into a series of ridges and depressions. Throughout large areas of Assam and Bengal, similar lurches were frequently seen after the earthquake of 1897. The rice-fields in these districts are carefully levelled so as to allow of their being flooded to a shallow and uniform depth. They were thrown into gentle undulations, the difference of level between crest and hollow being as much as 2 or 3 feet. In the earthquake of 1886, the ground about 16 miles from Charleston was compressed into similar undulations, the railway-lines being bent in a vertical plane so as to conform with the undulations*.

110. Buckling of Railway-Lines. The buckling of railway-lines usually, however, takes place in a horizontal plane, and is always a test of compression. The nature of the buckling depends on whether the compression be that of the solid crust or of the surface alluvium. In the former case (as in the Baluchistan earthquake of 1892), the buckling is uncompensated by extension elsewhere. In the latter case, which is by far the more common, buckling at any point is always accompanied by an opening of the rails, as a rule at distances of between a hundred yards and a quarter of a mile. Important as it is from the engineer's point of view, the buckling of railway-lines is only referred to here owing to its significance as an indication of the local compression of the ground†.

Earth-Fissures

111. In the central tract of all great earthquakes, the ground is extensively fissured. In the Charleston earthquake of 1886, fissures were a conspicuous and almost universal phenomenon within an area of about 600 sq. miles, and throughout a much larger region they occurred in great numbers, though with less continuity and at wider intervals.

Earth-fissures are easily distinguished from rifts or horizontal fault-movements by their great number, their inferior length

* Lawson, pp. 386, 401; Oldham, pp. 95, 194–195; Dutton, pp. 289, 291.

† C. Davison, *Geol. Mag.*, 1893, pp. 356–360; J. Milne, *Seis. Journ.*, vol. 1, 1893, p. 134; Dutton, pp. 283–295; Oldham, pp. 97–99.

and the inconstancy of their direction. They may be divided into five classes: (i) rift-fissures, due to the differential movement along a fault or rift; (ii) bluff-fissures, formed on hill-sides or in horizontal ground near trenches; (iii) plain fissures formed in horizontal ground but unconnected with trenches; (iv) hill-foot fissures, formed in alluvial plains at the foot of hills; and (v) fault-block fissures, consisting as a rule of two parallel fissures with depressed ground between. Rift-fissures have been described in Chapter V (sect. 76), and will not be referred to further in this chapter, as they are not secondary effects of the shock*.

112. Bluff-fissures. Fissures on hill-sides are merely small incipient landslips, and both owe their origin to the same cause. The fissures run parallel or nearly so to the strike of the subjacent rock-surface, and there is usually a slip of the ground on the downward side of the fissure. In the case of incipient earth-slumps, the fissure is horseshoe-shaped. One formed during the Andalusian earthquake of 1884 was about 2 miles long, from 10 to 50 feet wide, and of great depth. It was bordered by innumerable minor fissures, some parallel and others perpendicular to the main fissure. Houses in the enclosed tract were shifted as much as 30 yards within a month after the earthquake.

113. On more or less level ground, bluff-fissures occur near the banks of rivers, or the sloping sides of reservoirs or tanks, and on elevated roadways and river-embankments. They are seldom straight for any distance, following the curve of the river-banks, to which they are nearly always parallel. As a rule, they are arranged in concentric curves, being closest together near the river-banks, where the distance between successive fissures may be only a foot or two, but more frequently ranges from 10 to 15 feet. In length, they vary greatly. Some are only a few yards long, the majority attain a length of a few hundred feet, some even a few hundred yards, while a length of a mile is exceptional. They are often, however, arranged in series, overlapping one another *en échelon*, so that the total length of such a series mav be considerable. Their width is also variable. In the Charleston earthquake of 1886, they were seldom more than an inch wide except near river-banks. In

* Dutton, pp. 280–281.

other earthquakes, they may be 2 or 3 feet, or as much as 5 feet, wide. The depth is limited by that of the superficial layer of alluvium, and rarely, it would seem, exceeds 15 or 20 feet.

The origin of bluff-fissures is evident from their arrangement and distribution. They are clearly due to the settling of the ground near the unsupported margin of a trench or channel, and thus are merely incipient or undeveloped forms of land-slips *.

114. Fissures unconnected with Excavations. While the existence of a neighbouring trench is essential to the occurrence of bluff-fissures, other fissures are occasionally formed in flat alluvial ground far removed from any excavation, but only with earthquakes of great intensity. They differ from bluff-fissures in their greater magnitude and in the approximate constancy of their direction throughout a large area. During the New Madrid earthquake of 1812, many such fissures were formed in the alluvial plain of the Mississippi. In all, there was a tendency towards arrangement in parallel lines, the average direction being about N. 30° E. When the earth-lurches in this district took the form of long low undulations, the fissures were usually aligned in the same direction. In the Assam earthquake of 1897, the fissures were found in many cases to run parallel to raised roads or embankments and on either side of them.

There can be little doubt that the fissures here considered owe their origin to the visible waves which traversed the central district. (i) The formation of the fissures has been seen to coincide with the passage of the waves. During the New Madrid earthquake of 1812, the earth is said to have rolled in waves 3 feet high with visible depressions between, the waves finally breaking and leaving a series of parallel fissures. During the Assam earthquake of 1897, fissures were also seen to open as one wave had passed and to close up again as another arrived. (ii) As already noticed, some of the fissures of the New Madrid earthquake were parallel to the undulations of earth-lurches. Thus, though many fissures so formed may have closed up again,

* R. Mallet, *The Great Neapolitan Earthquake of* 1857, vol. 2, 1862, pp. 362–365; J. Milne, *Seis. Journ.*, vol. 1, 1893, p. 133; Dutton, pp. 280–281; C. S. Middlemiss, *Mem. Geol. Surv. India*, vol. 38, 1910, p. 122; Oldham, pp. 10, 13, 21; Lawson, pp. 401–402; Fuller, pp. 47–56.

others must have continued open, especially when the compression of the ground remained permanent in the form of earth-lurches *.

115. Hill-foot Fissures. During the Assam earthquake of 1897, fissures were formed at the foot of the Khasi and Garo hills wherever the alluvium of the plains extended up to the hill-side. It was found that the alluvium had separated from the hills and had dropped almost vertically through a distance of 1 to 5 feet, as at *bc* (Fig. 44), giving the appearance of a fault-scarp, except that it followed the windings of the hill-foot. Beyond this, a strip of the alluvium from 10 to 20 feet wide was slightly depressed, and, still farther, a band, *d*, slightly raised above the original level of the plain. Shortly after the

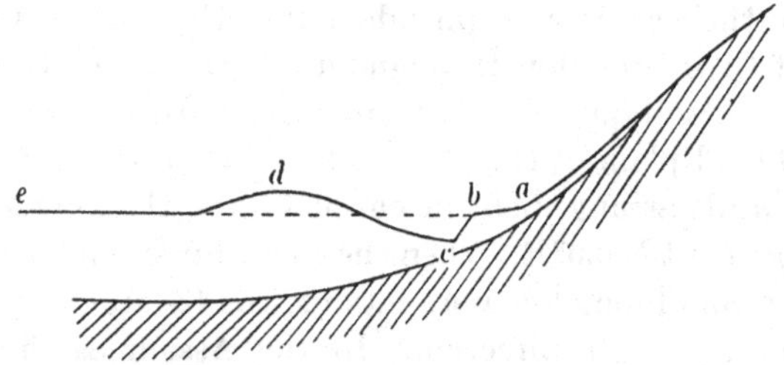

Fig. 44. Hill-foot fissure in the Assam earthquake of 1897.

earthquake, the varying height of the alluvium was rendered evident by the flooding of the plain for purposes of irrigation, the water covering the undisturbed surface but not the elevated strip. Oldham attributes the formation of hill-foot fissures to the repeated thrust of the hill and plain, the one against the other, during the shock, the alluvium lurching into a low ridge during the compression, while the drop and depression were formed during the return movements †.

116. Fault-Block Fissures. Fault-block or compound fissures have been observed in the central districts of the New Madrid earthquake of 1812 and the Assam earthquake of 1897, especially in those areas in which a layer of clayey alluvium overlies one of waterlogged sand. They consist of a pair of parallel fissures between which the ground has sunk (Fig. 45). Unlike

* Fuller, pp. 47–57; Oldham, pp. 20, 26, 89–90; R. Mallet, *Rep. Brit. Ass.*, 1850, p. 52.

† Oldham, pp. 92–93.

bluff-fissures, fault-block fissures are often straight for considerable distances. In some cases, they are so regularly formed that they might easily be mistaken for artificial trenches, the steep sides and flat bottoms resembling those of canal excavations. In the New Madrid earthquake, the parallelism of the fault-block fissures is generally very marked, groups of two to five or more long, straight and parallel canal-like depressions being not uncommon. The spaces between them are usually much greater than between ordinary fissures, amounting to several hundred feet and occasionally to half a mile. In length, they are also greater, the average being from 300 to 500 feet, though lengths of half a mile or more have been observed. In the Assam earthquake, the depth of the depressed region between the fissures was about a foot or 18 inches. In both earth-

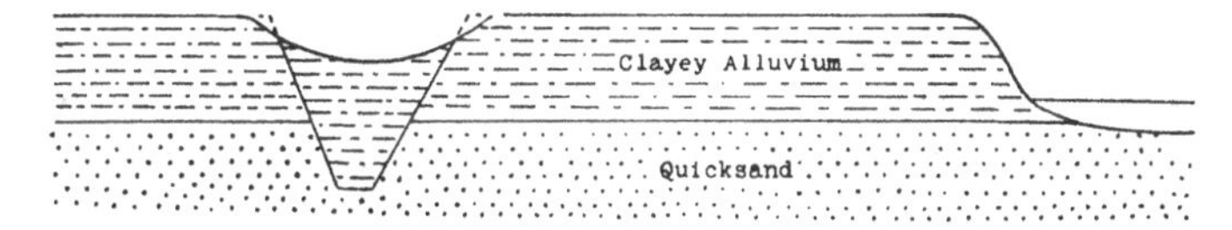

Fig. 45. Fault-block fissures in the New Madrid earthquake of 1812.

quakes, little or no sand was extruded through the fissures, and thus the undermining which allowed of the subsidence of the fault-block must have taken place by the extrusion of the quicksand through more distant fissures or by its creep into the neighbouring rivers*.

EFFECTS ON UNDERGROUND-WATER

117. Effects on Springs and Wells. The general effect of an earthquake is to raise the level of water in wells and to increase the flow from springs. The effect is, however, by no means uniform; for, with the same earthquake, there may be a diminished supply from some springs while others cease to flow. In some cases, the change of level in wells is of brief duration, passing away as soon as the earthquake is over; but, as a rule, it lasts from a few hours to a week or more. Before the Colchester earthquake of 1884, the water-level at the Colchester waterworks had been sinking. The shock, which was one of the strongest known in this country, caused a rise of 7 or 7½ feet.

* Oldham, pp. 92–93; Fuller, pp. 47–58.

The level gradually declined, but remained above the normal height for about six months.

In connexion with this subject, some interesting observations have been made by F. H. King on the level of the ground-water in a well in Wisconsin, U.S.A. The well is 140 feet from the nearest railway-line. The diagrams furnished by the gauge showed frequent sharp short-period curves denoting a rise in the level of the water, and these curves were found to be associated with the movement of trains past the well. The most marked rises were produced by heavily loaded trains which moved rather slowly. The diagram indicated a rapid but gradual rise of the water, followed by a slightly less rapid fall to the normal level, the movement beginning shortly after the passage of the train and amounting to about one-tenth of an inch. Somewhat similar observations have occasionally been made on the level of well-water during earthquakes.

There are thus two distinct movements in the underground water associated with earthquakes, (i) a somewhat sudden rise in the level of the water at the time of an earthquake, followed by a rapid return to the normal level, and (ii) a rise in the water, returning to the original level after the lapse of weeks or months. The causes of the movements are not entirely clear, but it is probable that the short-period variations are due to the transitory extrusion of the capillary water from the interstices of the soil; while the long-period variations are caused by a temporary widening or closing of the fissures through which the water circulates*.

118. Extrusion of Water from Fissures. The extrusion of water from fissures is a common phenomenon of great, and even moderately strong, earthquakes. It does not, however, take place from all fissures, but is usually confined to certain belts, chiefly no doubt to those in which the surface alluvium overlies a bed of water-bearing sand.

The water is often ejected with considerable force, carrying with it sand and other material from some depth. In several earthquakes, it has been seen to rise in continuous columns or

* R. Meldola and W. White, *East Anglian Earthquake of April 22nd*, 1884, pp. 155–162; F. H. King, *U. S. Department of Agriculture, Weather Bureau, Bull.* No. 5, 1892, pp. 67–69; C. E. de Rance, *Nature*, vol. 30, 1884, p. 81.

jets to heights of from 2 to 4 feet, while stray splashes, as indicated by the sand left on trees, have reached a height of 13 feet or more. The ejection continues for several minutes, and in some cases for hours, after the earthquake. The total amount of water extruded may therefore be very great. After the Charleston earthquake of 1886, every stream-bed, though usually dry in summer, carried off the ejected water. During the New Madrid earthquake of 1812, the water forced its way through the surface-deposits, "blowing up the earth with loud explosions." The quantity extruded was enormous, one tract many square miles in area being covered to a depth of 3 or 4 feet.

The extrusion of water and sand is due to the compression of the layer of water-bearing sand between the alluvium and the underlying beds. This compression may be effected partly by the shock itself, and, in some cases, partly by the upward thrust of the underlying beds. The continuance of the extrusion after the shock is over is probably due to the differential settling of the fissured alluvium at the surface*.

SAND-VENTS

119. The extrusion of water, accompanied by sand, is of two general types, (i) the more common form of violent ejection in jets described in the preceding section, and (ii) the quiet extrusion of water. With each type is associated its peculiar form of sand-vent; with the former, the sand-craters common in all great earthquakes, and with the latter the rare form of sand-blows, sand-sloughs, etc., manifested during the New Madrid earthquake of 1812.

120. Sand-Craters. A typical sand-crater is shown in Fig. 46, which represents one of those formed during the Assam earthquake of 1897. It consists of a saucer-shaped crater of sand, with its rim slightly raised above the original level of the ground.

Sand-craters are of all sizes from about a foot to 20 feet or more in diameter, and several feet in depth. They occur only in the central area of the earthquake, and in those parts of it in which there is a water-bearing layer of sand at a short distance

* R. Mallet and T. Oldham, *Quart. Journ. Geol. Soc.*, vol. 28, 1872, pp. 259–260, 266–269; Oldham, pp. 13, 15, 20, 101–104; Lawson, p. 403; Dutton, pp. 281–282, 289; Fuller, p. 76; R. Meldola and W. White, *East Anglian Earthquake of April 22nd*, 1884, p. 77.

below the surface. In some places, they are close together, almost overlapping; in others, they are several yards apart. After the Assam earthquake of 1897, 52 craters were counted in a strip of land near Maimansingh containing about one-eighth of an acre. The amount of sand ejected from the craters is often considerable. In the central district of the Charleston earthquake, many acres were covered with sand, which was 2 feet or more in thickness near the orifices and thinned out towards the margins. In this earthquake, the sand, as a rule, came from a few feet only below the surface; in the Californian earthquake

Fig. 46. Sand-vent in the Assam earthquake of 1897.

of 1906, it was in one place the same as that pierced by wells at a depth of 80 feet.

The formation of sand-craters is due to the extrusion of water and sand, as described in the last section. By the rapid outflow of the water, the fissures at one or more points are worn into round channels of considerable size, which are still further enlarged at the surface by the back-rush of the water into the fissure. As the force of the current begins to slacken, a circular mound of sand is deposited round the orifice, within the rim of which a hollow is formed by the water rushing downwards. Gradually, the orifice becomes choked with sand, through which the water filters back; and thus the bottom of the crater, instead of narrowing almost to a point, is broad and nearly flat, though slightly hollowed*.

* Dutton, pp. 281, 284; Oldham, p. 20; Lawson, p. 403; R. Mallet and T. Oldham, *Quart. Journ. Geol. Soc.*, vol. 28, 1872, pp. 266–269.

121. Sand-Blows and Sand-Sloughs. The typical sand-blow of the New Mad id earthquake is a nearly circular patch of sand, from 15 to 18 feet in diameter, and 3 to 6 inches high, with a low rounded profile with concave slopes, but without any trace of a central depression. The large sand-blows are 100 feet or more in diameter and about a foot in height. Some are elongated in form, about 200 feet long and 25 to 50 feet wide. They are confined to the low alluvial lands of the Mississippi basin, but are absent from the immediate neighbourhood of the river. In places, they merged into one another, so that the whole surface was covered with a continuous sheet of sand.

Sand-sloughs are low, somewhat ill-defined, ridges of sand parallel to one another, and alternating with shallow troughs in which water is collected in long narrow pools. The ridges and depressions, as a rule, are only a few inches in height or depth. Sand-sloughs are always found on low ground, but very rarely near the Mississippi.

Both sand-blows and sand-sloughs are formed by the quiet extrusion of water and sand, caused probably by the unequal settling of the alluvial deposits into the water-bearing stratum below. Their absence from the higher lands appears to be due to the greater thickness of the deposits above the quicksand in those regions; and their absence from the neighbourhood of the Mississippi to the water and quicksands flowing laterally into the river*.

Rise of River-Channels, etc.

122. During the Assam earthquake of 1897, river-channels, tanks, wells, etc., were filled up over a large area, partly perhaps by the abundant outpouring of sand, but chiefly by the forcing-up of the bottom. Trenches or excavations of any kind are places of special weakness in the surface layer, and the sand underlying it would be forced up readily through such openings. Many river-channels from 15 to 20 feet deep were almost completely obliterated in this manner, so that the rivers afterwards flowed on shallow sandy beds.

Another effect, due to the same changes, was a sudden rise in the rivers of from 2 to 10 feet, which took place immediately

* Fuller, pp. 79–89.

after the earthquake, but passed off in a day or two. At Gauhati, for instance, the Brahmaputra rose more than 7½ feet within three-quarters of an hour of the earthquake, but returned to its normal level in 2½ days. At Goalpara, the first rise in the same river was not less than 10 feet.

This rise may have been partly due to the great quantity of water extruded from the sand-vents, but it was too sudden to be entirely explained in this manner. It was probably caused by the forcing-up of the river-bed, more in some parts than in others, so that the water was partially ponded back by barriers. Being composed of sand, such barriers would be rapidly scoured away and the river would return to its normal level*.

Effects of Earthquakes on Glaciers

123. The effects of earthquakes on glaciers have been studied in the case of one series of earthquakes, those of Alaska in 1899. To the north of the epicentral district of Yakutat Bay lie the lofty St Elias and other ranges, from which many glaciers descend towards the coast. The effects of the earthquakes on these glaciers are of two kinds.

(i) An immediate effect of the earthquakes was the shattering of the glacier-ice over a wide area, a shedding of icebergs from the glaciers which descend into the sea, and a consequent recession of the glaciers. Even in Glacier Bay (150 miles from Yakutat Bay), the front of the well-known Muir Glacier was so shattered that the inlet into which the glacier flowed was inaccessible to steamships for eight years. During the 13 years from 1894 to 1907, the total retreat of the Muir Glacier was 8½ miles, and of the Grand Pacific Glacier 8 miles, the retreat in both cases being largely, but perhaps not entirely, due to the earthquakes of 1899.

(ii) A second and far more important effect of the earthquakes was due to the vast snow-avalanches that fell from the Alaskan mountains and accumulated in the glacier reservoirs. In 1899, all the Yakutat Bay glaciers were retreating, and this retreat continued for several years longer, being no doubt aggravated by the shedding of icebergs at the time of the earthquakes. Then, one by one, the glaciers began to advance, the shortest

* Oldham, pp. 104–108.

first at some time before 1905, and then in turn those of increasing length from 1905 onwards, the longest of all being still unaffected in 1910. The changes undergone were similar in all the glaciers. The advance was abrupt and spasmodic; the surface, previously smooth and easily traversed, was transformed into a wilderness of pinnacles and crevasses; the glaciers thickened at their lower ends; and finally, after advancing several hundred yards in ten months or less, the effects of the increased supply of snow were spent and the glaciers returned rapidly to a stagnant condition*.

* R. S. Tarr and L. Martin, *U. S. Geol. Surv.*, Prof. Paper No. 69, 1912, pp. 51–61.

CHAPTER VIII

POSITION OF THE SEISMIC FOCUS

124. The main subject of all earthquake-investigations is the position of the seismic focus. If the focus were a point, the elements required would be the latitude and longitude of the epicentre and the depth of the focus below the surface. In many catalogues, the latitude and longitude of the epicentre are given, the reference in such cases being to the position of a point which may be regarded as the centre of the epicentral tract or of a point which coincides approximately with that in which the shock attained its greatest strength. Even Mallet, to whom we are indebted for the term seismic focus, realised that the dimensions of the focus must be considerable. His investigation of the Neapolitan earthquake of 1857—the first attempt to grapple seriously with the problem before us—led to the conclusion that the focus was a fissure 10 miles long in a horizontal direction, and in depth varying from $4\frac{3}{4}$ to $8\frac{1}{4}$ miles.

The realisation that the focus possesses finite magnitude increases the complexity of the problem. We have not only to determine the mean position of the epicentre; we have also to ascertain the form and dimensions of the epicentral tract. To a great extent, both are problems within our present grasp, and the method of their solution will now be described. With regard to the depth of the focus, our methods are less accurate and, consequently, our knowledge is uncertain. Some of the methods which have been suggested will be explained below. The present chapter will be confined to the case of earthquakes in which the epicentral tract can be studied or is near at hand. The determination of the origin of distant and unfelt earthquakes will be considered in the next chapter (sects. 162–166).

Position of the Epicentre

125. In a great earthquake, in which the central tract is crushed and faulted, the determination of the epicentre is usually a simple matter. The dislocation of the ground and the scarp of the faulted rock are more or less permanent traces of a portion of the epicentral tract. The whole of the area it may be more difficult to outline. The distribution of the after-shocks, as in the

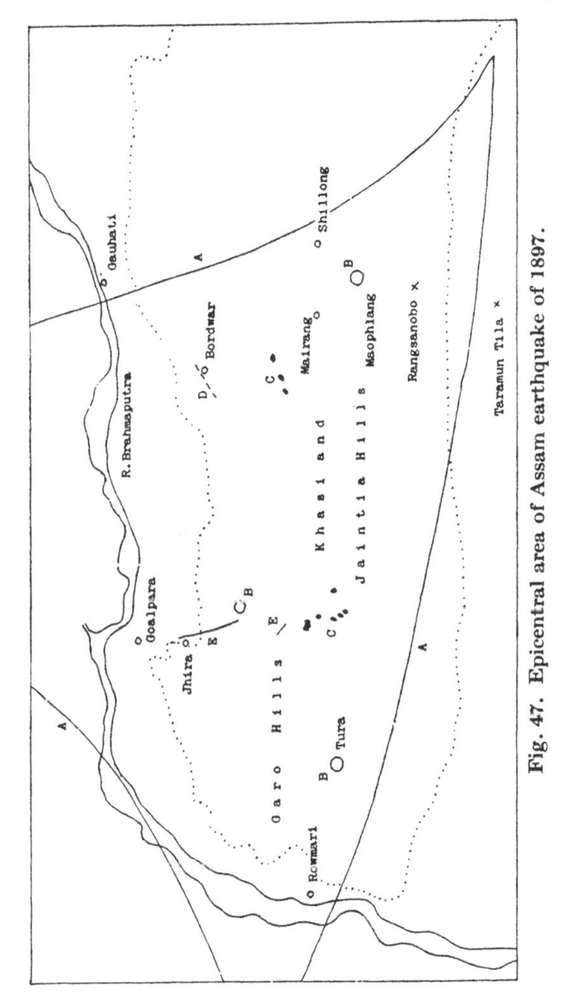

Fig. 47. Epicentral area of Assam earthquake of 1897.

Mino-Owari earthquake of 1891, may indicate a fault-displacement of which no signs are left on the surface (sects. 208, 215). In a complex earthquake, like the Assam earthquake of 1897, Oldham used various lines of evidence—the course of the fault-scarps and rock-fissures, the folding of the crust, and the distribution of the after-shocks—and assigned to the epicentral area the form shown in Fig. 47, a vast region 200 miles in length and 6000 sq. miles in area*.

In earthquakes unaccompanied by fault-displacements, we have to rely on other evidence, the methods which have been suggested depending on observations of the time, the direction of the shock, and its intensity, respectively.

126. Methods depending on the Time. The methods which depend on observations of the time of occurrence at five or more stations need not be described in any detail. They have been used in many earthquakes, but the results obtained are untrustworthy, because (i) it has never been found possible to obtain sufficiently accurate records of the time; and (ii) even if the observations were accurate to the nearest second, different phases of a movement lasting for many seconds or several minutes might be timed in different places.

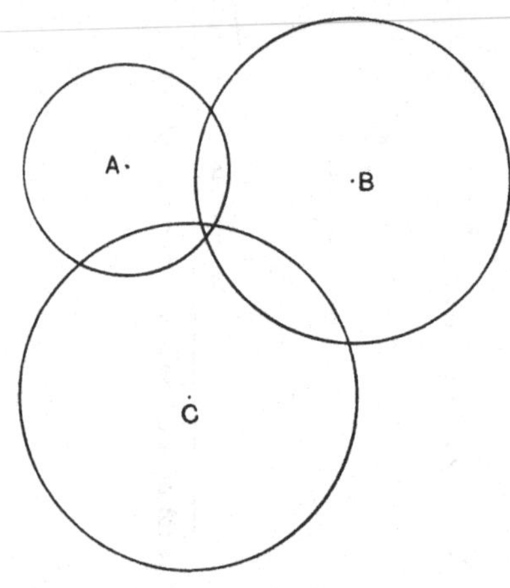

Fig. 48. Omori's method of determining the epicentre.

Omori's method, which depends on the duration of the preliminary tremor at three or more stations, is free from these objections. If y be the duration in seconds of the preliminary tremor at a place distant x kms. from the focus, Omori has shown (sect. 60), that, in earthquakes in which x is less than 1000, x and y satisfy the equation

$$x = 7{\cdot}27y + 38.$$

If the durations of the preliminary tremor be, say, 40, 65 and 72 seconds at the stations A, B and C, respectively (Fig. 48), the distances of the focus from them must be 330, 510 and 550 kms. Assuming that the depth of the focus is small compared with these distances, it follows that the epicentre must lie on each of the three circles described with A, B and C as centres

* Oldham, pp. 164–179.

and the above distances as radii, and therefore at or near the point in which the circles intersect. In practice, the circles do not pass through a point, but form a small triangle, the centre of which may be regarded as giving the position of the epicentre.

The method is used in Japan for determining the epicentres of the submarine earthquakes which originate off the east coast of that country*.

127. Method depending on the Direction. The second method —that depending on observations of the direction of the shock— is the well-known method used by Robert Mallet, though suggested by John Michell in 1760. Mallet assumed that, during the passage of the wave, each particle of rock moves in a closed curve the longer axis of which is directed towards the focus, and, in practice, he regarded the curve as a straight line. He inferred that monuments, gate-posts and loose objects would fall, or be projected, in the direction of the wave-path at each place, and that fissures in the walls of buildings—and he laid stress on the necessity of the buildings being simple and symmetrical in shape—would be formed along perpendicular lines. As such wave-paths would intersect in the epicentre, two lines of direction would be sufficient if the epicentre were a point, while, if otherwise, a larger number would intersect two and two at different points within the epicentre and might thus plot out roughly its form and magnitude.

Though simple in form, the method is by no means easy to apply. The first aspect of a ruined city, as Mallet found in studying the Neapolitan earthquake of 1857, is one of bewildering complexity. Houses are overthrown in every conceivable direction. There seems at first sight to be no governing law in their fall. It is only, as he remarks, by first gaining some commanding point, and by viewing the place as a whole, that any prevailing direction seems to stand out clearly from the surrounding confusion. If Mallet had measured the bearing of every suitable fissure and of the fall of every object, and had taken the mean of all the observations, he would probably have obtained a very close approximation to the principal direction of the shock, as Omori did with the fallen stone-lamps at Tokyo (sect. 58). The method he adopted was apparently

* F. Omori, *Publ. Eq. Inv. Com.*, No. 21, 1905, pp. 31–33.

to form a first mental impression of the mean direction and then to measure particular lines which seemed to agree most closely with that impression.

The results of Mallet's investigation of the Neapolitan earthquake are shown in Fig. 49. The line drawn from each place represents the direction of the shock at that place. As will be seen, these lines are not absolutely concurrent, though they show a marked tendency to converge towards the village of

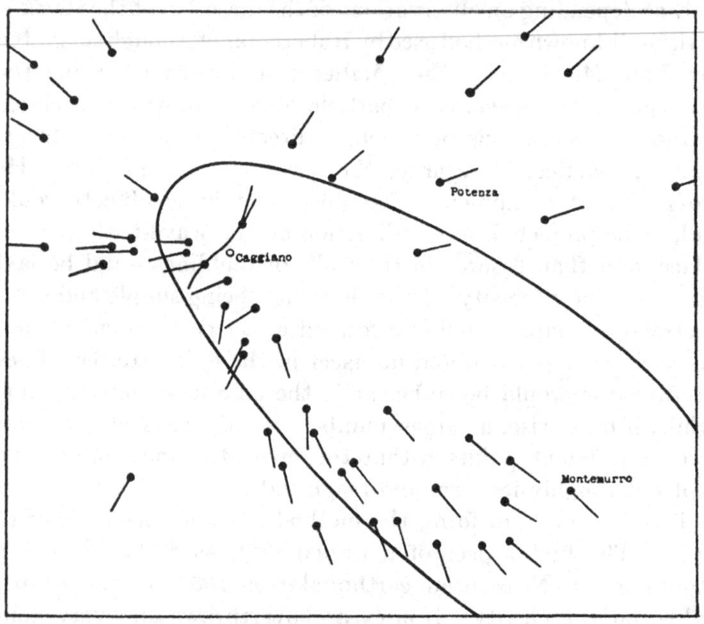

Fig. 49. Mallet's method of determining the epicentre of the Neapolitan earthquake of 1857.

Caggiano. They intersect, indeed, at various points within an area indicated by the curve on the map—a line which Mallet regarded as the surface-trace of a fissure about 10 miles long, within which the earthquake probably originated.

The method of directions, which Mallet used with such skill in this disastrous earthquake, has lately fallen into undeserved neglect. The extreme variability of the direction in any place cannot be overlooked, and solitary observations are not only valueless, but misleading. But, if the method be modified by the substitution of the mean of a large number of measurements for a few scattered observations, it may still be found a useful instrument in the investigation of a destructive earthquake (sects. 58, 59)*.

128. Method depending on the Intensity. The first method depending on observations of the time, even if it could be applied with accuracy, requires a large disturbed area for its use. For

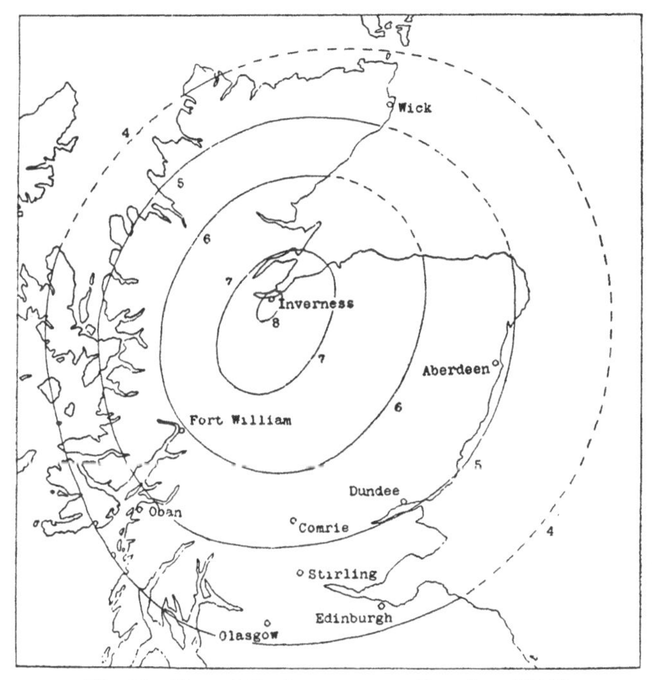

Fig. 50. Map of the Inverness earthquake of 1901.

the method of directions, a strong shock is also essential. The chief advantage of the third method is that it can be applied in earthquakes of every degree of strength. In violent shocks, the area of ruined towns and villages usually encloses, if it does not mark out, the epicentral area. In the weakest shocks, the disturbed area is itself so small that any place at which it is felt may be regarded without much error as coincident with the epicentre.

* R. Mallet, *The Great Neapolitan Earthquake of* 1857, vol. 2, 1862, pp. 235–247.

In these cases, the determination of the epicentre is only approximate. Greater accuracy is attained by the construction of a series of isoseismal lines, the centre of the innermost curve being either close to, or within, the epicentral area. Thus, as shown in Fig. 50, the epicentre of the Inverness earthquake of 1901 lies about 3 miles south-west of the city of Inverness; the principal epicentre of the Derby earthquake of 1904 is about $1\frac{1}{2}$ miles east of Ashbourne (Fig. 51).

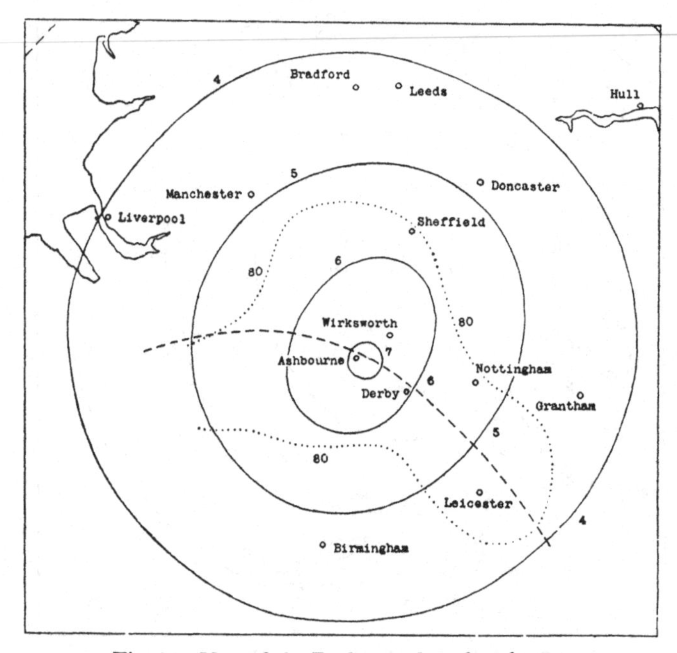

Fig. 51. Map of the Derby earthquake of 1904.

In the map of Fig. 51, the centre of the innermost curve is excentric with respect to the next surrounding isoseismal. In this earthquake, the shock consisted of two distinct parts. A shock of this type is known as a *twin* earthquake, and it will be seen later (sect. 243) that such earthquakes owe their origin to nearly simultaneous impulses in two detached foci. The more important focus is that which corresponds with the centre of the innermost isoseismal. In a few earthquakes, as in the Doncaster earthquake of 1905 and the Kangra earthquake of

1905 (Fig. 52), it is possible to trace the isoseismal lines surrounding both epicentres and therefore to assign their positions.

129. Form and Position of the Seismic Focus. A carefully drawn series of isoseismal lines reveals, however, much more than the position of the epicentre. As will be seen from the examples here given, the curves are usually elongated in form, and the distances between successive isoseismals are unequal on the two sides of the axes. These are necessary results of a long

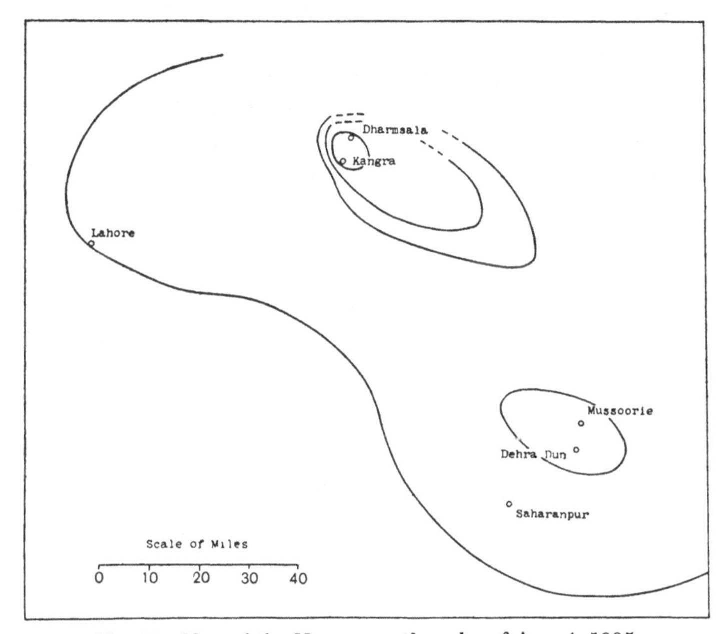

Fig. 52. Map of the Kangra earthquake of Apr. 4, 1905.

seismic focus inclined to the horizon, such as a portion of a fault-surface within which one rock-mass slips over the other. From the forms and relative position of the isoseismal lines, we may deduce the elements of the fault, the displacement along which results in a given shock. (i) The direction of the fault is parallel, or nearly parallel, to the longer axes of the isoseismal lines. (ii) The fault hades towards the side on which the inner isoseismals are the farther apart. (iii) The fault-line passes close to the centre of the innermost isoseismal, and lies on that side of it from which the fault hades. (iv) The difference between the lengths of the axes of the innermost isoseismal gives a rough measure of the depth of the seismic focus.

The Inverness earthquake of 1901 (Fig. 50) may be taken as an example of these methods. The innermost isoseismal (of intensity 8, Rossi-Forel scale) is 12 miles long and 7 miles wide, the longer axis being directed N. 33° E. The next isoseismal (intensity 7) is 53½ miles long, 35 miles wide, with its longer axis directed N. 32° E. The distance between the isoseismals is 9 miles on the north-west side and 14 miles on the south-east side. From these measurements, it may be inferred that the mean direction of the originating fault is about N. 33° E., that the fault hades to the south-east, that the fault-line passes a short distance on the north-west side of the centre of the isoseismal 8, and that the length of the principal part of the focus was at least 5 miles. Now, the epicentral area is crossed by the great northern boundary fault of the Highland district, the line of which passes at a distance of somewhat less than a mile (Fig. 97) on the north-west side of the centre of the isoseismal 8. The fault hades to the south-east and, in the neighbourhood of the epicentre, its mean direction is about N. 35° E. It thus satisfied all the conditions implied by the seismic evidence*.

DEPTH OF THE SEISMIC FOCUS

130. While the position of the epicentre can be determined with some approach to accuracy, much less confidence can be felt in our estimates of the depth of the focus. That the upper margin of the focus may sometimes coincide with the surface is obvious—the formation of a fault-scarp is evidence of this. It is clear, also, that, in most cases, the mean depth of the focus is not considerable, that it must be measured in terms of a few miles rather than of many miles. (i) The limited area over which many earthquakes are felt shows that, in such cases, the seismic focus is shallow. As will be seen later (sect. 139), the more rapidly the intensity of a shock declines outward from the epicentre, the less must be the depth of the focus. (ii) Earthquakes can only be produced by fault-slipping in that outer

* C. Davison, *Beitr. zur Geoph.*, vol. 9, 1908, pp. 220–224; *Quart. Journ. Geol. Soc.*, vol. 58, 1902, pp. 377–397.

portion of the crust which is so rigid that the rock under strain will break rather than bend. (iii) In those earthquakes which are traced to movements along known faults, the epicentre lies on the downthrow side of the fault and at distances from the fault-line not exceeding 1 or 2 miles*.

131. Methods depending on the Time. If the seismic focus were a point and the shock of brief duration, accurate time-observations at five or more stations would give the position

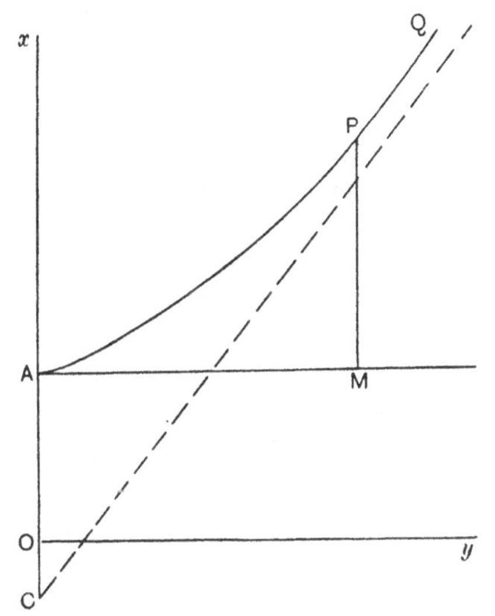

Fig. 53. Von Seebach's time-curve.

of the epicentre and the depth of the seismic focus. Such estimates of the depth have been made for several earthquakes†.

* For instance, in the Bolton earthquake of Feb. 10, 1889 (Fig. 29), the epicentre was 2 miles from the Irwell Valley fault; the epicentres of the Inverness earthquake of Sep. 18, 1901, and its numerous accessory shocks lay between $\frac{1}{10}$ mile and $1\frac{1}{2}$ miles from the Great Glen fault (Fig. 97); the epicentre of the Malvern earthquake of 1907 was about a mile from the boundary fault of the Malvern Hills; the epicentres of the numerous Ochil earthquakes of 1900–1914 lie within 2 miles of the great fault which runs along the southern margin of the Ochil Hills. In the case of the Bolton earthquake of 1889, the hade of the fault is known, namely, 28°; as the point of maximum intensity of the shock is about $4\frac{1}{2}$ miles from the fault-line, it follows that the depth of the focus was about $3\frac{3}{4}$ miles.

† For instance, Hogben estimates the depth of the focus in the New Zealand earthquake of 1888 to be about 24 miles, in that of 1890 to be between 20 and 30 miles, and in that of 1891 to be probably a little less than 10 miles (*Trans. New Zeal. Inst.*, 1890, pp. 471, 477; 1892, p. 365).

But little or no reliance can be placed on the results, for the focus is always of some size, time-observations are seldom accurate to the nearest minute, and in any case may refer to different phases of the motion.

Von Seebach has suggested a graphical method of dealing with observations of the time. In the accompanying diagram (Fig. 53), O represents the focus, A the epicentre, and AM a horizontal line along the surface. The distance of any place of observation is represented by a length AM measured along this line. From the other end M, a perpendicular line MP is drawn of a length proportional to the difference between the observed times at the place and the epicentre, that is, to the difference between the distances OM and OA. The points corresponding to the other end P of this and similar lines would, if all the measurements were exact, lie on the hyperbola APQ, from the form of which it is possible to calculate the depth of the seismic focus*. In practice, the points so plotted fall into groups on either side of a curve which is drawn so as to pass as closely to them as possible.

Von Seebach's method has been used in the investigation of a few earthquakes, with the following results†:

Earthquake	Authority	Mean depth in miles, about
Rhenish, 1846	von Seebach	24
Sillein, 1858	,,	16
Mid-German, 1872	,,	12
Herzogenrath, 1873	von Lasaulx	7
,, 1877	,,	17

* Taking the vertical and horizontal lines through O as axes of x and y, let x, y be the coordinates of P, and c the depth of the focus. Let

$$MP = \lambda\,(OM - OA),$$

where λ is a constant. Then

$$x - c = \lambda\,(\sqrt{c^2 + y^2} - c),$$

or

$$x^2 - \lambda^2 y^2 - 2\,(1 - \lambda)\,cx + (1 - 2\lambda)\,c^2 = 0,$$

which is the equation of an hyperbola. The equation of the asymptote in Fig. 53 is $x - \lambda y = c\,(1 - \lambda)$, the line passing through the seismic focus when $\lambda = 1$, in which case the equation of the curve becomes $x^2 - y^2 = c^2$.

† K. von Seebach, *Das mitteldeutsche Erdbeben vom 6. Marz 1872* (Leipzig, 1873), pp. 132–175; A. von Lasaulx, *Das Erdbeben von Herzogenrath am 23. October 1873* (Bonn, 1874), and *Das Erdbeben von Herzogenrath am 24. Juni 1877* (Bonn, 1878).

Von Seebach assumed that the velocity of the earth-waves is constant at all depths below the surface. If it should vary with the depth, the time-curve, as A. Schmidt has shown, would be no longer hyperbolic, but concave upwards at and near the epicentre and afterwards convex. The tangent to the curve at the point of inflexion intersects the vertical through the focus at a point above the focus, and thus the depth of this point would give an inferior limit to the depth of the focus. Galitzin, making use of good observations of the time at stations near the epicentre, thus found the depth of the focus of the South German earthquake of Nov. 16, 1911, to be 5·9 miles, with a maximum error of 2·4 miles*.

132. Another method depending on time-observations has been suggested by Omori. If y be the duration in seconds of the preliminary tremor at a place near the focus and distant x kms. from it, Omori has shown (sect. 60) that x and y in ordinary earthquakes satisfy the equation

$$x = 7{\cdot}42y.$$

Thus, if the position of the epicentre be known, and if the distance of the focus be calculated from this equation, the depth of the focus is easily ascertained.

Omori has applied this method to the earthquakes which accompanied the eruptions of the Asama-yama in central Japan. In 1911, these earthquakes were registered at Ashino-taira, the horizontal distance of which from the crater is 5 kms. The mean duration of the preliminary tremors was ·96 second, giving 7·2 kms. for the distance of the focus. It follows that the mean origin must be about 6 kms. below the summit of the mountain or about 4 kms. (2½ miles) below its base. In 1911 and 1912, the earthquakes were registered at Yuno-taira, the horizontal and vertical distances of which from the crater are 2·3 and ·6 kms. The mean duration of the preliminary tremors being ·64 second, it follows that the mean focus was 4·8 kms. from the observatory, or 4·8 kms. below the summit of the mountain or about 3·0 kms. (nearly 2 miles) below its base†.

* A. Schmidt, *Jahr. des Vereins für vaterl. Naturk. in Württemb.*, 1888, pp. 248–270; 1900, pp. 200–232 (abstract in *Nature*, vol. 52, 1895, pp. 631–633); B. Galitzin, *Compt. Rend. des Séances de la Com. Sis. Perm.*, vol. 5, 1913, p. 335.

† F. Omori, *Bull. Eq. Inv. Com.*, vol. 6, 1912, pp. 127–128, 238.

133. Method depending on the Direction. This method, which is due to Mallet, depends on the assumptions (i) that it is possible to determine the direction of the wave-path at one or more points on the surface, and (ii) that the direction of the wave-path at any point coincides with that of the focus. If the focus were a point and if the position of the epicentre were known, one measurement of the direction of the wave-path would be sufficient to determine the depth of the focus. If, on the other hand, the focus were of finite magnitude, the intersections of the wave-paths at different pairs of points might give some idea of the depths of the upper and lower margins of the focus.

Mallet sought to determine the direction of the wave-path by measuring the inclination to the vertical of fissures in fractured walls. He chose in all cases large walls, composed of bricks or short-bedded stones, and containing very few windows and doors. For instance, in the cathedral at Potenza, he found, from a series of nearly parallel fissures, that the mean angle of emergence was 23° 7′. The distance of Potenza from the epicentre at Caggiano being 17 miles, the depth of the focus below the surface of the ground would be 17 tan 23° 7′, or 7¼ miles, or about 6¾ miles below the level of the sea. Mallet also measured the angle of emergence at 25 other places, the farthest being Salerno, distant 40 miles from the epicentre, at which the angle was 11½°. The calculated depths corresponding to the different angles of emergence lie between 3 and a little over 9 miles with a mean depth of about 6½ miles*.

Now, if the waves proceeded from such depths as these, there can be no doubt that, in their passage to the surface, they would traverse rocks of varying density and elasticity, and that at every bounding surface they would be deflected from their previous course. If the true direction of the wave-path at Potenza were 5° more or less than that given above, the corresponding depth of the focus below sea-level would be 8½ or 5 miles. If the observation at Salerno were subject to the same error, the depth might be 11¼ or 4, instead of 6½, miles.

* R. Mallet, *The Great Neapolitan Earthquake of* 1857, vol. 2, 1862, pp. 248–251.

In the following table* are given the approximate depths of the foci of seven earthquakes, obtained by Mallet's direction-method:

Earthquake	Authority	Mean depth in miles, about
Neapolitan, 1857	Mallet	$6\frac{1}{2}$
Ischian, 1881	Johnston-Lavis	$\frac{1}{3}$
„ 1883	„ „	$\frac{1}{3}$
„ „	Mercalli	$\frac{3}{4}$
Andalusian, 1884	Taramelli and Mercalli	$7\frac{1}{2}$
Kashmir, 1885	Jones	$7\frac{1}{2}$
Riviera, 1887	Mercalli	$10\frac{3}{4}$
Verny (Turkestan), 1887	Mouchketow	$6\frac{1}{2}$

134. In his investigation of the Yokohama earthquake of 1880, Milne applied the direction-method to determine the depth of the focus, but estimated the angle of emergence from the vertical and horizontal components of the motion as registered by seismographs. The results of different observations varied within wide limits, but Milne considered the most probable depth to be from $1\frac{1}{2}$ to 5 miles. In two other Japanese earthquakes (Feb. 7 and Apr. 30, 1897) Omori and Hirata estimated from seismographic records at Miyako that the depths of the foci were 5·6 and 9·3 miles. The same objection of course applies to these observations as to Mallet's. Possibly the angles of emergence in the Miyako records are determined with greater accuracy than those from the directions of fissures, but they depend on the further assumption that there was at the time no tilting of the ground†.

* R. Mallet, *Great Neapolitan Earthquake of* 1857, vol. 2, 1862, pp. 240–251; H. J. Johnston-Lavis, *Monograph of the Earthquakes of Ischia*, 1885, pp. 71–74; G. Mercalli, *L' isola d'Ischia ed il terremoto del 28 luglio* 1883, pp. 133–134; T. Taramelli and G. Mercalli, *Mem. R. Accad. dei Lincei*, vol. 3, 1886, pp. 69–71; E. J. Jones, *Rec. Geol. Surv. India*, vol. 18, 1885, pp. 224–225; G. Mercalli, *Ann. dell' Uff. Centr. di Meteor. e Geodin.*, vol. 8, 1888, pp. 246–248; J. V. Mouchketow, *Mém. du Com. Géol.*, vol. 10, p. 149. In addition to the above are the following high estimates: 30 miles for the Cachar earthquake of 1869 (T. Oldham, *Mem. Geol. Surv. India*, vol. 19, 1882, pp. 66–71), and 50 miles for the Bengal earthquake of 1885 (C. S. Middlemiss, *Rec. Geol. Surv. India*, vol. 18, 1885, p. 211).

† J. Milne, *Trans. Seis. Soc. Japan*, vol. 1, part 2, 1880, p. 104; F. Omori and K. Hirata, *Journ. Coll. Sci.*, Imp. Univ. Tokyo, vol. 11, 1899, pp. 194–195.

135. Methods depending on Intensity. Two methods have been suggested, one by Mallet and the other by Dutton. Both depend on the rate at which the intensity declines outward from the epicentre. In both, it is assumed that the intensity of the shock varies inversely as the square of the distance from the focus, and that there is no loss of energy, such as by reflection and refraction of the waves at the bounding surfaces of different strata.

(i) Mallet regards the power of a shock to overthrow buildings as proportional to the horizontal component of the intensity in the direction of the wave-path, which is zero at the epicentre

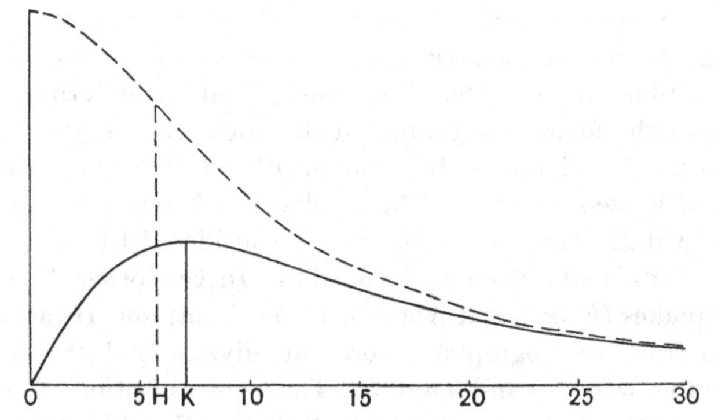

Fig. 54. Diagram illustrating Mallet's and Dutton's methods of determining the depth of the focus from variations of intensity.

and a maximum along a circle which he calls the meizoseismal line. The depth of the focus is equal to the radius of the circle multiplied by $\sqrt{2}$ or 1·414....

(ii) Dutton considers only the intensity of the shock without regard to direction. This declines uniformly outwards from the epicentre, but its rate of decline is greatest at a distance from the epicentre, which, multiplied by $\sqrt{3}$ or 1·732...., gives the depth of the focus.

In Fig. 54, the continuous line represents the horizontal component of the intensity, and the broken line the intensity, at different distances from the epicentre. The focus in each case is supposed to be at a depth of 10 miles. The maximum horizontal component would be greatest at the distance represented by *OK* and the maximum rate of decline of the intensity at the distance represented by *OH* *.

136. Both methods are open to the same objections:

(i) The difficulty of fixing the exact distances at which the maxima are attained, and this applies especially to Dutton's method;

(ii) The assumption that the focus is either a point or a region symmetrical in all directions;

(iii) The assumption that energy is not lost during the progress of the wave. For instance, if the depth of the focus were 10 miles, the maximum horizontal component would always be attained at a distance of about 7 miles, and the maximum rate of decline of intensity at a distance of about 6 miles, from the epicentre. Yet it is conceivable that an earthquake might originate at the above depth and not be felt at a distance of 6 or 7 miles from the epicentre, perhaps not felt at the surface at all. If the methods were unexceptionable in other respects, it would be more nearly correct to say that they give a lower limit to the depth of the seismic focus.

137. Mallet's method was used by Middlemiss in his investigation of the Bengal earthquake of 1885; the radius of the meizoseismal line was estimated at 74 miles, leading to the clearly excessive depth of 104 miles.

Dutton's method has been applied to several earthquakes with the following results†:

* If c be the depth of the focus and x the distance of any place from the epicentre, the intensity of the shock at that place may be represented by $\mu/(c^2 + x^2)$, and its horizontal component by $\mu x/(c^2 + x^2)^{\frac{3}{2}}$. The rate of decline of intensity is represented by the differential coefficient of the first expression with respect to x or $-2\mu x/(c^2 + x^2)^2$. Equating to zero the differential coefficients of the second and third expressions, we obtain the equations

$$c^2 - 2x^2 = 0 \text{ and } c^2 - 3x^2 = 0,$$

leading to the results given above. (R. Mallet, *Rep. Brit. Ass.*, 1858, pp. 101–102; Dutton, pp. 313–317.)

† C. S. Middlemiss, *Rec. Geol. Surv. India*, vol. 18, 1885, p. 211; Dutton, pp. 317–320; D. Eginitis, *Ann. de Géogr.* (Paris), 1895; S. Arcidiacono, *Ann. dell' Uff. Centr. Met. e Geod.*, vol. 16, 1895, p. 7; A. Cavasino, *Boll. Soc. Sis. Ital.*, vol. 18, 1914, pp. 433–435; C. S. Middlemiss, *Mem. Geol. Surv. India*, vol. 38, 1910, pp. 331–334; E. Oddone, *Boll. Soc. Sis. Ital.*, vol. 19, 1915, pp. 190–194; M. Stuart, *Rec. Geol. Surv. India*, vol. 49, 1918, pp. 187–188.

Earthquake	Authority	Mean depth in miles, about
Charleston, 1886 (two foci)	Dutton	12 and 8
Constantinople, 1894	Eginitis	21
Syracuse, 1895	Arcidiacono	$4\frac{1}{2}$
Marsica (Italy), 1904	Cavasino	$4\frac{1}{3}$
Kangra, 1905	Middlemiss	between 21 and 40
Marsica, 1915	Oddone	6
Srimangal (Assam), 1918	Stuart	8 or 9

138. The methods used for determining the depth of the focus have been described at some length, chiefly on account of the interest of the problem. The most accurate results are probably those of Johnston-Lavis for the earthquakes of Ischia, and, next to these, may be placed Omori's for the earthquakes of the Asama-yama. Estimates greater than 10 or 12 miles are based on scanty evidence. Of the remainder, the utmost that can be said is that they indicate, not the depth of the focus, but simply its order of magnitude.

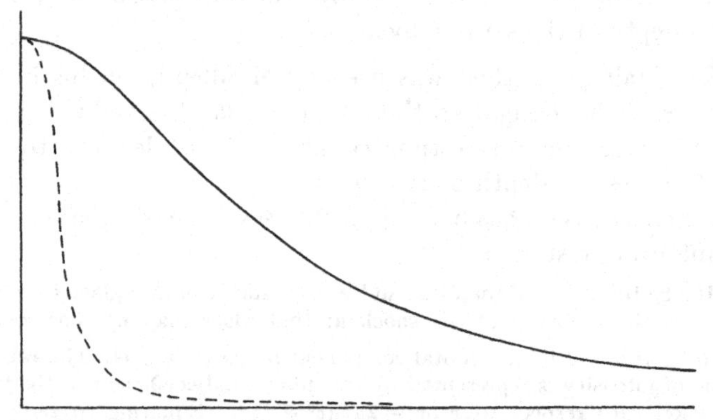

Fig. 55. Diagram illustrating the relative depths of the foci of two earthquakes determined by the more or less rapid variations of intensity.

139. The principle involved in Dutton's method (sect. 135) gives, however, a test of the relative depths of the foci of two earthquakes, which is often useful. The curves in Fig. 55 represent the intensities of the shock at different distances from the epicentre, the continuous line corresponding to a focus 2 miles in depth, the broken line to one at a depth of one-quarter of a mile. They are drawn on the assumptions that the intensity of the shock at any point varies inversely as the square of its distance from the focus, and that the impulses in the two foci are such that the intensity at the epicentre is the same in each case. Thus, the more rapid the decline outwards in the intensity of the shock, the less is the depth of the focus.

For instance, many Etnean earthquakes (sect. 221) destroy every building within a small zone and yet disturb areas of only 100 or 200 square miles. In Great Britain, the strongest earthquakes are just capable of causing slight damage to buildings, yet the average area disturbed by them is more than 65,000 sq. miles. We may infer with confidence that the foci of the Etnean earthquakes are very much shallower than those of British earthquakes.

CHAPTER IX

PROPAGATION OF EARTHQUAKE-WAVES

140. When a disturbance takes place in an isotropic solid—that is, a body which has the same elastic properties in all directions about a point—two principal types of waves are propagated outwards from the source. In one, the particles vibrate along lines normal to the wave-front; in the other, in the plane perpendicular to this direction. During the passage of the former waves, the solid is alternately compressed and rarefied; during the passage of the latter waves, it is distorted. The waves are thus known as *longitudinal* or *condensational* or *compressional* waves and *transverse* or *distortional* waves, respectively.

If ρ be the density of the solid, k the resistance to compression, measured by the pressure required to produce unit contraction, and n the modulus of rigidity, measured by the stress required to produce unit shearing-strain, and if $m = k + \frac{4}{3}n$, the velocities of the waves are given respectively by the expressions $\sqrt{(m/\rho)}$ and $\sqrt{(n/\rho)}$. Since m is greater than n, it follows that the velocity of the condensational wave always exceeds that of the distortional wave*.

* No branch of seismology has attracted more attention recently than the subject of this chapter. Among the large number of memoirs devoted to it, the following may be mentioned as especially worthy of attention:

1. Geiger, L. and B. Gutenberg. Über Erdbebenwellen. *Nach. der K. Gesell. der Wissen. zu Göttingen, Math.-phys. Klasse,* 1912, pp. 623–675.
2. Knott, C. G. (1). Reflexion and refraction of elastic waves, with seismological applications. *Phil. Mag.*, vol. 48, 1899, pp. 64–97, 567–569.
3. —— (2). *The Physics of Earthquake Phenomena* (Oxford Univ. Press), 1908, pp. 156–258.
4. —— (3). The propagation of earthquake waves through the earth, and connected problems. *Proc. Roy. Soc. Edin.*, vol. 39, 1919, pp. 157–208.
5. Milne, J. *Reports of the Seismological Committee of the British Association*, 1896–1913.
6. Oldham, R. D. (1). On the propagation of earthquake motion to great distances. *Phil. Trans.*, 1900 A, pp. 135–174.
7. —— (2). The constitution of the interior of the earth, as revealed by earthquakes. *Quart. Journ. Geol. Soc.*, vol. 62, 1906, pp. 456–473.
8. Omori, F. Horizontal pendulum observations of earthquakes at Tokyo.

REFLECTION AND REFRACTION OF EARTHQUAKE-WAVES

141. In passing from one medium to another possessing different wave-moduli or different densities, both condensational and distortional waves are modified. Each may be resolved into four new waves—a reflected condensational, a reflected distortional, a refracted condensational and a refracted distortional, wave—though, under certain conditions, one or more of the waves may be absent. The number and types of such subsidiary waves depend partly on the angle of incidence of the earthquake-wave, and partly on the relative values of the elastic moduli and densities of the two media*.

Publ. Eq. Inv. Com., No. 5, 1901, pp. 1–82; No. 6, 1901, pp. 1–181; No. 13, 1903, pp. 1–142; No. 21, 1905, pp. 9–102.

9. Reid, H. F. Instrumental records of the [Californian] earthquake. *The Californian Earthquake of April* 18, 1906 (edited by A. C. Lawson), vol. 2, 1910, pp. 59–142.

10. Turner, H. H. *Reports of the Seismological Committee of the British Association*, from 1914.

11. Walker, G. W. *Modern Seismology* (Longmans), 1913, pp. 37–82.

12. Wiechert, E. and K. Zoeppritz. Über Erdbebenwellen. *Nach. der K. Gesell. der Wissen. zu Göttingen, Math.-phys. Klasse*, 1907, pp. 1–135.

Experiments on the determination of the elastic constants of various rocks are described by H. Nagaoka in *Publ. Eq. Inv. Com.*, No. 4, 1900, pp. 47–67; and by S. Kusakabe in *Journ. Coll. Sci.*, Imp. Univ. Tokyo, vol. 19, 1903, art. 6, and vol. 20, 1905, arts. 9, 10; also in *Publ. Eq. Inv. Com.*, No. 17, 1904, pp. 1–48; No. 22 B, 1906, pp. 27–49.

* Let ρ, ρ' be the densities of the two media, m, n and m', n' their elastic moduli; let θ be the angle of incidence (and reflection) of a condensational wave travelling in the former medium, θ' its angle of refraction, and ϕ, ϕ' the angles of reflection and refraction of the resulting distortional waves. These quantities are connected by the following equations:

$$\frac{m}{\rho}\operatorname{cosec}^2\theta = \frac{n}{\rho}\operatorname{cosec}^2\phi = \frac{m'}{\rho'}\operatorname{cosec}^2\theta' = \frac{n'}{\rho'}\operatorname{cosec}^2\phi';$$

equations which express the fact that the paths followed by the waves are such that they are traversed in the shortest possible times.

Since m is greater than n, it follows that $\operatorname{cosec}^2\phi$ is greater than $\operatorname{cosec}^2\theta$ and therefore greater than unity, that is, the equation always gives a possible value of ϕ, and there will always be a reflected distortional wave.

If m/ρ be greater than m'/ρ' (and therefore greater than n'/ρ'), the above equations, for the same reason, give possible values of θ' and ϕ'; that is, there will always be refracted waves of both types.

If, however, m/ρ be less than m'/ρ' though greater than n'/ρ', there will as before be a refracted distortional wave. There will also be a refracted condensational wave, if $\frac{m}{\rho}\operatorname{cosec}^2\theta \Big/ \frac{m'}{\rho'}$ be greater than unity; but, if θ should exceed the value for which this expression is equal to unity, then

142. The energies of the reflected and refracted condensational and distortional waves have been calculated by Knott for various media*. A simple case is that in which the first medium is rock (of density 3, rigidity $1{\cdot}5 \times 10^{11}$, Poisson ratio ·25) and the second water, the refracted distortional wave being of course absent. Taking the total energy of all three waves as unity, the curves in Figs. 56 and 57 represent, for varying angles of incidence, the energies of the subsidiary waves, the incident wave (travelling in rock) being condensational in Fig. 56 and distortional in Fig. 57. In each figure, the angle of incidence is measured along the horizontal line, the continuous line represents the energy of the reflected condensational wave, the broken line that of the reflected distortional wave, and the dotted line that of the refracted condensational wave.

When the incident wave is condensational, it is seen from Fig. 56 that (i) the energy of the reflected condensational wave diminishes at first as the angle of incidence increases, is practically zero when the angle lies between 50° and 80°, and increases rapidly to unity when the angle becomes 90°; (ii) the energy of the reflected distortional wave increases at first with the angle of incidence, being about three-quarters of the total energy when the angle lies between 45° and 80°, and decreases rapidly to zero when the angle becomes 90°; and (iii) the energy of the refracted condensational wave is two-fifths of the total energy at normal incidence, and gradually diminishes as the angle of incidence increases, until it vanishes when the angle is 90°. Two particular cases, when the angles of incidence are respectively 30° and 70°, are further illustrated in Fig. 58. The line on the

cosec θ' is less than unity or θ' becomes imaginary, that is, there is no refracted condensational wave.

Lastly, if m/ρ be less than n'/ρ' (and therefore less than m'/ρ'), there will be a refracted condensational wave if $\frac{m}{\rho} \operatorname{cosec}^2 \theta \Big/ \frac{m'}{\rho'}$ be greater than unity; but, if θ should exceed the value for which this expression is equal to unity, the refracted condensational wave is absent. Similarly, there will be a refracted distortional wave if $\frac{m}{\rho} \operatorname{cosec}^2 \theta \Big/ \frac{n'}{\rho'}$ be greater than unity; but, if θ should exceed the value for which this expression is equal to unity, then the refracted distortional wave also disappears, and the incident wave is replaced entirely by condensational and distortional reflected waves.

* Knott (1), pp. 64–97.

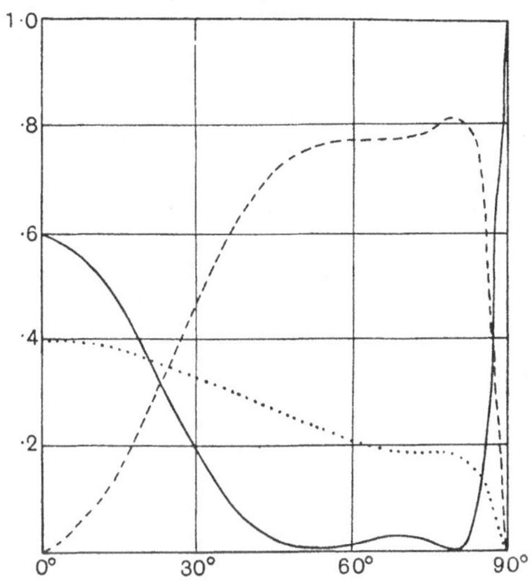

Fig. 56. Diagram representing the energies of the reflected and refracted waves, rock to water, incident wave condensational.

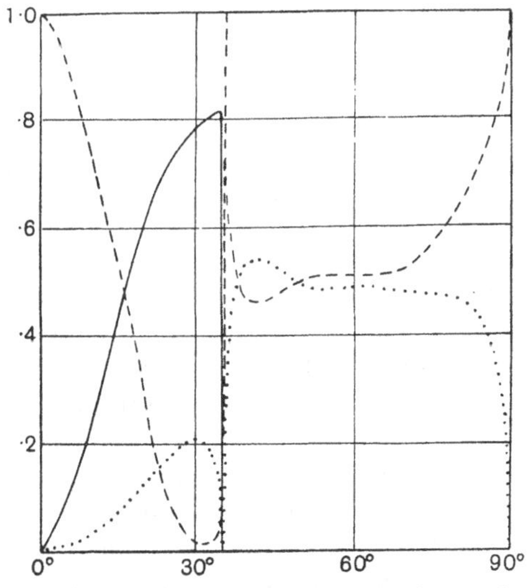

Fig. 57. Diagram representing the energies of the reflected and refracted waves, rock to water, incident wave distortional.

left of the normal represents the direction and energy of the incident condensational wave. Of the lines on the right of the normal, the continuous line represents the direction and energy of the reflected condensational wave, the broken line those of the reflected distortional wave, and the dotted line those of the refracted condensational wave.

When the incident wave is distortional, it is seen from Fig. 57 that (i) the energy of the reflected condensational wave, starting from zero, increases rapidly with the angle of incidence until just before the angle attains the value 35° 6′, when the energy suddenly decreases to zero, after which this wave ceases to exist; (ii) the energy of the reflected distortional wave rapidly

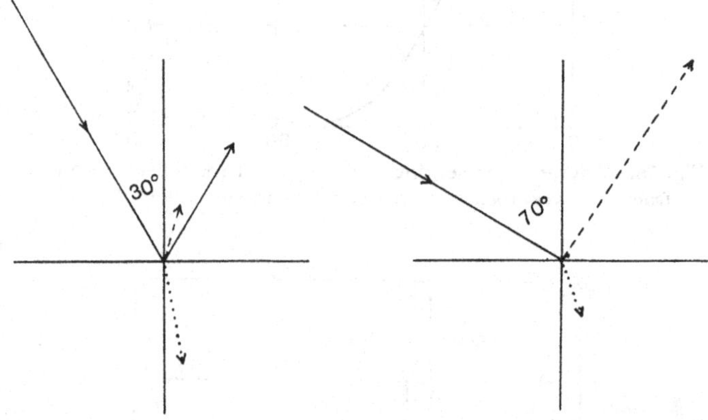

Fig. 58. Diagram representing the directions and energies of the reflected and refracted waves, rock to water, incident wave condensational.

falls as the angle of incidence increases, is very nearly zero just before the angle reaches the above critical value, when it increases suddenly to unity, after which it rapidly decreases to slightly less than half the total energy, and then increases once more to unity when the angle is 90°; and (iii) the energy of the refracted condensational wave is small while the angle of incidence is less than the critical value, after which it becomes about half the total energy until the angle exceeds 80°, and then rapidly decreases to zero when the angle is 90°. Three particular cases, when the angles of incidence are respectively 21° 47′, 35° 6′ and 70°, are further illustrated in Fig. 59. The line on the left of the normal represents the direction and energy of the incident distortional wave. The lines on the right of the normal represent the directions and energies of the subsidiary waves as in Fig. 58.

Knott's calculations also give the angles of refraction corresponding to different values of the angles of incidence. He shows that the angle of refraction cannot exceed 23° when the

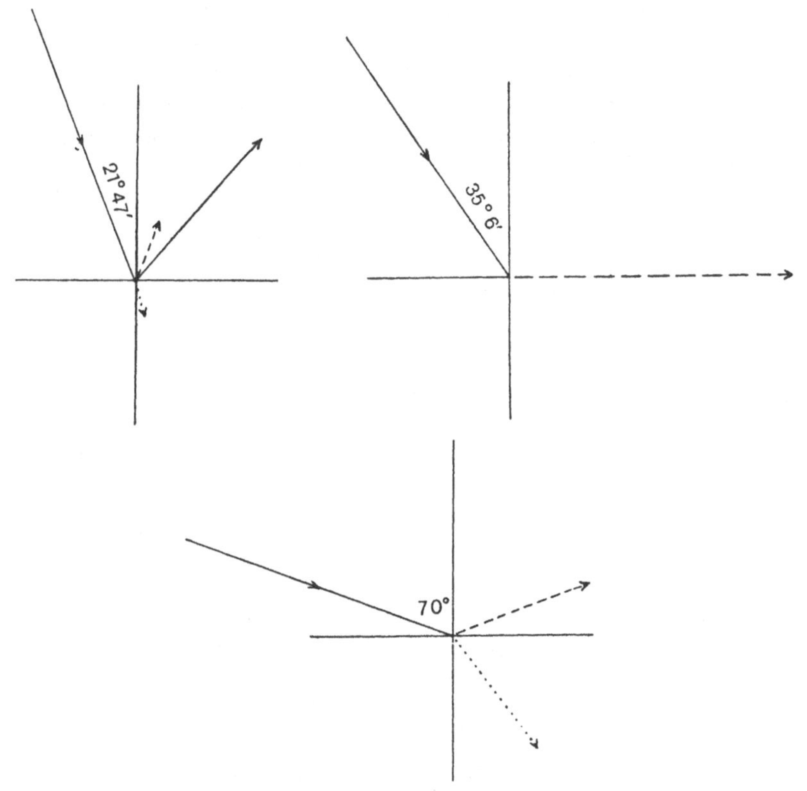

Fig. 59. Diagram representing the directions and energies of the reflected and refracted waves, rock to water, incident wave distortional.

incident wave is condensational, or 42° when it is distortional. Thus, the refracted condensational wave must as a rule travel almost directly upwards to the surface of the sea; and this explains the nature of seaquakes, as if the ships in which they are felt were bumping over rocks. When the second medium is air instead of water, the angle of refraction cannot exceed 5°, and thus the earthquake-sound waves should appear to come directly from below.

143. When the media are both solids the curves corresponding to Figs. 56 and 57 are more complicated. Knott considers several cases. For instance, when the elastic constants are the same for both solids and the densities are as 1 to 2, an incident wave in the less dense medium gives rise to distortional waves, reflected and refracted, of small energy; about 90 per cent. of the initial energy is concentrated in the reflected condensational wave until the angle of incidence is nearly 60°, after which the refracted condensational wave increases in prominence until it possesses rather more than half the initial energy when the angle of incidence is about 84°. If the incident wave be in the denser medium, the reflected condensational wave absorbs more than three-fourths of the total energy until the angle of incidence is about 45°, after which it ceases to exist. The energies of the other three waves, small for small angles of incidence, suddenly increase in the neighbourhood of the above critical angle, to amounts between one-quarter and one-half of the total energy; but, when the angle of incidence exceeds 60°, the energies of the distortional waves are both small, while the energy of the refracted condensational waves increases to more than 93 per cent. of the total when the angle of incidence is about 84°.

144. With other values, and especially with widely differing values, of the elastic constants and densities, the waves which are relatively unimportant in the above example may become more prominent. There is no need to consider particular cases, the main conclusion being obvious, namely, that, in the heterogeneous outer crust of the earth, the seismic waves, however simple they may be initially, must be so split up by frequent reflection and refraction at the bounding surfaces of different rocks that, in the neighbourhood of the epicentre, the condensational and distortional waves can never be completely separated. To the continual concurrence of waves of both types, we must assign the confused character of the records of near earthquakes illustrated in the third chapter. Further applications of Knott's interesting results will be seen in sects. 158, 159, 161.

EARTHQUAKE-MOTION AT GREAT DISTANCES FROM THE ORIGIN

145. The undulations of a great earthquake are propagated far beyond the limits of the disturbed area, unfelt owing to the smallness of their amplitudes or the length of their periods. Some early observations of these unfelt waves may be noticed first. The Lisbon earthquake of Nov. 1, 1755, produced oscillations or seiches, several feet in range, in pools and lakes in Scotland, Norway and Sweden, Lake Wener in the latter country being about 1750 miles from Lisbon. A great earthquake occurred at Iquique on May 9, 1877; 78 minutes later, unusual oscillations were observed in the bubble of a level at Pulkowa, more than 8000 miles away. The Andalusian earthquake of Dec. 25, 1884, disturbed magnetographs at Lisbon, Parc Saint-Maur (near Paris), Greenwich and Wilhelmshaven. The Riviera earthquake of Feb. 23, 1887, was recorded by magnetographs at the same places, as well as at Montsouris (near Paris), Nantes, Pola, Vienna, Brussels, Utrecht and Kew. During the night of Sep. 10, 1893, the measurements of C. V. Boys at Oxford on the density of the earth were interrupted by the vibrations of a comparatively slight earthquake in Roumania. During the year 1889, a horizontal pendulum erected by E. von Rebeur-Paschwitz at Potsdam to measure the lunar disturbance of gravity recorded the Tokyo earthquake of Apr. 17, the Verny (Central Asia) earthquake of July 12, the Kumamoto earthquake of July 28, and the Patras (Greece) earthquake of Aug. 25 *.

In these cases, the instruments affected were not specially designed for the registration of earthquakes. They indicated, however, the lines on which suitable seismographs should be constructed, and several of those described in the second chapter (sects. 25, 26, 28) date from the closing years of the last century. In the same year (1893), E. von Rebeur-Paschwitz and J. Milne suggested a world-wide distribution of stations for the observation of distant earthquakes, and, two years later, under the auspices of the Seismological Committee of the British Association, Milne instituted the system which has added so much to our knowledge of the unfelt earth-waves †.

* M. Baratta (*Riv. Geogr. Ital.*, 1897) has collected accounts of similar disturbances, especially of magnetic needles.

† Rebeur-Paschwitz's suggestion led ultimately to the foundation in 1901 of the International Seismological Association with its head-quarters at Strasbourg, the useful career of which ended in 1914.

146. Examples of Earthquake-Records. The diagrams reproduced in Figs. 60–63 are typical seismograms of distant earthquakes, as given by different forms of seismographs. Fig. 60 is the record of the Californian earthquake of Apr. 18, 1906, at Edinburgh; Fig. 61 that of the Kangra earthquake of Apr. 4, 1905, at Birmingham; Fig. 62 that of an earthquake in Asia Minor on Feb. 9, 1909, at Pulkowa; and Fig. 63 that of an earthquake in the Pacific Ocean (near the Bonin Islands) on Nov. 24, 1914, at Bidston. Figs. 60 and 61 were registered by undamped instruments; the former, by a Milne seismograph, is given for its historical interest; the latter, by an Omori horizontal pendulum, to show the influence of a free pendulum on the nature of the record. Figs. 62 and 63 were provided by a Galitzin

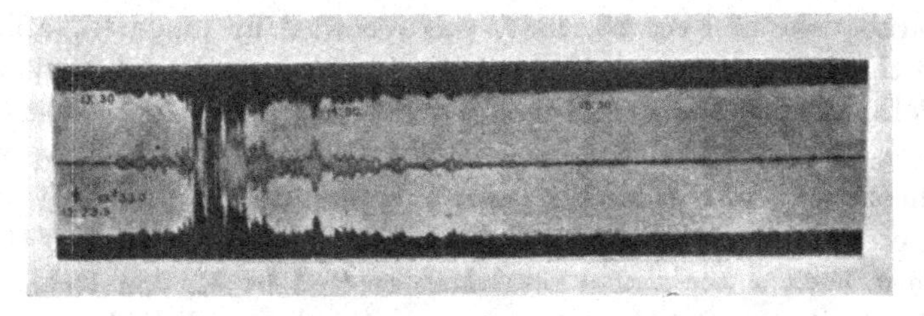

Fig. 60. Seismogram of the Californian earthquake of Apr. 18, 1906, at Edinburgh.

seismograph and Milne-Shaw seismograph respectively, and, as both instruments contain aperiodic pendulums, the resulting diagrams represent with the greatest accuracy the true motion of the ground.

147. Phases of Motion. Confining our attention to Figs. 62 and 63, it will be seen at once that the whole movement is divisible into well-marked phases, the beginnings of which are shown by the arrows *P*, *S* and *L*.

(i) The first phase consists of vibrations that are so weak that they sometimes escape record. If the previous trace be undisturbed by micro-tremors, the first movement is a small but sudden displacement from the straight line, followed by rapid but irregular vibrations, each lasting one or more seconds. In this phase, there are occasionally sudden reinforcements of the motion, such as that indicated by the arrow *PR*, to which reference will be made afterwards (sects. 158, 159). The vibrations of this phase are called the *first preliminary tremors* or *primary waves*.

(ii) The second phase is usually marked by a considerable increase in amplitude (as at *S*, Figs. 62, 63), the first vibration being succeeded by irregular movements, with an amplitude which may increase to a maximum and then on the whole decrease. In this phase, also, sudden reinforcements appear, such as that indicated by the arrow *SR*, which will be considered in sects. 158, 159. The vibrations of this phase constitute the *second preliminary tremors* or *secondary waves*.

(iii) As the irregularities of the second phase die away, they merge into the waves of long period which constitute the third phase. As a rule, the movement is both larger and more regular than in the preceding phases. The first few undulations are, however, irregular and with periods of sometimes as much as 40 seconds. These are followed by a series of regular harmonic vibrations, with periods of from 12 to 20 seconds, some of which may attain the largest amplitude exhibited during the whole motion. After this maximum phase is over, the undulations become less regular, though among them particular periods, such as 12 or 18 seconds, may prevail. Then gradually the movement dies down until, after the lapse of an hour or more, the trace becomes once more steady. The vibrations of this phase are known as the *long waves* or *large waves*. The maximum part of this phase is sometimes called the *principal portion* of the movement (subdivided by Omori into the initial phase, the slow-period phase and the quick-period phase), while the concluding undulations are known as the *end portion* or *coda*. Some time later, usually after this phase has ended, a group of waves, similar to those of the principal portion but of very small amplitude, may be seen, and occasionally, with very strong earthquakes, a second group of such waves. These will be considered more fully in sect. 160*.

148. The initial vibration of each phase is sometimes denoted by the letter *P*, *S* or *L*. A second meaning is, however, given

* When the seismograph is not damped, the movements recorded (as will be seen in Fig. 61) are somewhat different from those described above. The same three phases are clearly discernible, but, in each phase, the vibrations are more regular in period and more uniform in amplitude, and the undulations of the principal portion, the period of which approaches that of the proper oscillations of the pendulum, become unduly prominent.

to the same letters, namely, the times at which the respective phases begin. It is convenient to make use of both interpretations, as the double meaning gives rise to no ambiguity. It will be seen that the exact determination of the times *P*, *S* and *L* (and especially of *P* and *S*) at different stations is a problem of the greatest importance.

Of the three epochs, the first, *P*, when clearly defined, can be determined with the greatest accuracy. If the previous trace be free from micro-tremors, it can be measured to the nearest second. The beginning of the *S* wave, though always of larger amplitude, may be less sharply marked, for it is superposed on the concluding tremors of the primary waves. Of the three epochs, that of the *L* wave is usually ascertained with least accuracy; the waves are readily seen, but the exact instant when the secondary waves merge into the long waves is not easy to define.

It is only with a high magnifying power that it is possible to determine the epoch *P* at a considerable distance from the origin. For instance, the Kingston earthquake of Jan. 14, 1907, was recorded at three stations close together in the United States. The seismograph at Washington, with a magnifying power of 25, recorded the first preliminary tremors; the record at Cheltenham (magnifying power 10) began with the second preliminary tremors; that at Baltimore (magnifying power 6) showed only the principal part.

149. Time-Curves of the Earthquake-Phases. When the position of the epicentre is definitely known, the rates at which the initial waves of the different phases travel can be represented by a series of curves. As a rule, the distances in kilometres of stations measured along a great circle from the origin are measured along the horizontal axis, and the times at which the waves reach the stations by the lengths of lines perpendicular to that axis. The curves are drawn through or, as a rule, near the terminal points of the series of lines, due weight being given to observations of exceptional accuracy or to many independent and closely agreeing observations. They are known as the *time-curves* or *hodographs** of the earthquakes, and have been con-

* The word "hodograph" has already been adopted for an entirely different curve in dynamics. It is, however, frequently employed, and, in its seismological applications, is free from doubt.

structed for many individual earthquakes*. As they differ but little from the curves given in Fig. 64, it is unnecessary to reproduce a typical series of time-curves.

150. The forms of the time-curves for individual earthquakes

Degrees	*P* secs.	*S* secs.	*S–P* secs.	Degrees	*P* secs.	*S* secs.	*S–P* secs.	Degrees	*P* secs.	*S* secs.	*S–P* secs.	Degrees	*P* secs.	*S* secs.	*S–P* secs.
1	15	28	13	31	398	711	313	61	619	1116	497	91	801	1464	663
2	31	55	24	32	407	728	321	62	625	1128	503	92	807	1475	668
3	47	83	36	33	416	744	328	63	632	1141	509	93	812	1485	673
4	62	110	48	34	425	760	335	64	638	1153	515	94	818	1496	678
5	77	137	60	35	433	775	342	65	645	1165	520	95	823	1506	683
6	92	164	72	36	442	790	348	66	651	1177	526	96	829	1516	687
7	106	190	84	37	450	804	354	67	658	1190	532	97	834	1526	692
8	121	217	96	38	458	818	360	68	664	1202	538	98	840	1536	696
9	136	243	107	39	466	832	366	69	671	1214	543	99	845	1546	701
10	150	269	119	40	475	847	372	70	677	1226	549	100	851	1556	705
11	164	294	130	41	483	861	378	71	683	1238	555	101	855	1565	710
12	179	319	140	42	491	875	384	72	690	1250	560	102	860	1575	715
13	193	344	151	43	498	888	390	73	696	1262	566	103	865	1584	719
14	206	368	162	44	506	902	396	74	702	1274	572	104	870	1593	723
15	219	392	173	45	513	915	402	75	709	1286	577	105	874	1602	728
16	232	415	183	46	520	928	408	76	715	1297	582	106	879	1612	733
17	245	438	193	47	527	941	414	77	721	1309	588	107	884	1621	737
18	257	460	203	48	534	954	420	78	727	1320	593	108	888	1630	742
19	269	482	213	49	540	966	426	79	733	1332	599	109	893	1639	746
20	281	503	222	50	547	979	432	80	739	1343	604	110	897	1648	751
21	293	524	231	51	553	991	438	81	745	1355	610	111	902	1657	755
22	305	545	240	52	560	1004	444	82	750	1366	616	112	907	1666	759
23	317	565	248	53	566	1016	450	83	756	1377	621	113	911	1674	763
24	328	584	256	54	573	1029	456	84	762	1388	626	114	916	1682	766
25	338	603	265	55	579	1041	462	85	768	1399	631	115	920	1690	770
26	348	622	274	56	586	1054	468	86	773	1410	637	116	925	1698	773
27	358	641	283	57	592	1066	474	87	779	1421	642	117	929	1706	777
28	368	659	291	58	599	1079	480	88	785	1432	647	118	934	1714	780
29	378	677	299	59	605	1091	486	89	790	1443	653	119	938	1722	784
30	388	694	306	60	612	1103	491	90	796	1454	658	120	942	1729	787

vary within certain narrow limits, the variations depending partly on the depth of the focus, partly perhaps on the paths traversed by the waves. At present, attention is being concentrated on the average time-curves for all earthquakes. First constructed by Milne in 1899, these curves were year after year

* The time-curves of the following earthquakes may be mentioned as examples: Assam earthquake of June 12, 1897 (Oldham, plate 39); Kangra earthquake of Apr. 4, 1905 (F. Omori, *Publ. Eq. Inv. Com.*, No. 24, 1907, plates 3–7); Californian earthquake of Apr. 18, 1906 (Reid, plate 2); Marsica earthquake of Jan. 13, 1915 (G. Agamennone, *Boll. Soc. Sis. Ital.*, vol. 22, 1919, plates 1, 2).

corrected by him as fresh and more accurate observations became available, and tables were compiled giving the times of transit for different arcual distances measured in degrees. Milne's

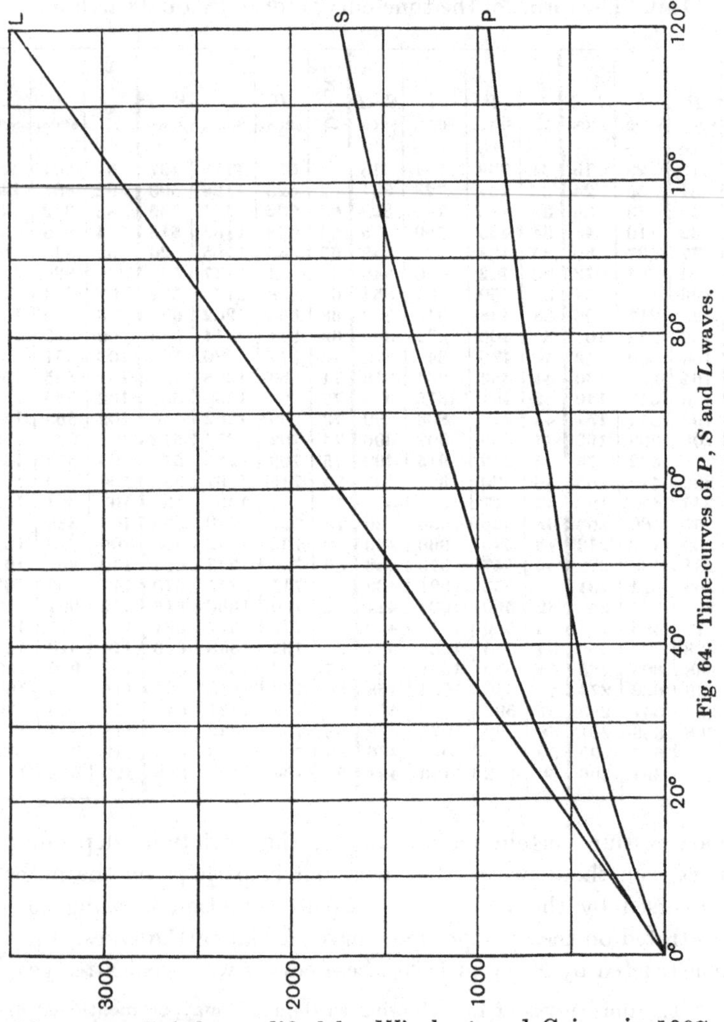

Fig. 64. Time-curves of P, S and L waves.

tables were slightly modified by Wiechert and Geiger in 1907. The latest tables are those given by H. H. Turner, depending mainly on the observations collected for the Seismological Committee of the British Association. For the first 120 degrees, they are here reproduced, the curves P and S in Fig. 64 representing graphically the variations in the times of the initial primary and secondary waves with the distance. The line L is the corresponding curve for the long waves, as determined by the observations collected by O. Klotz*. A study of the table and time-curves leads to some important conclusions.

151. (i) **Velocities of Earth-Waves.** Denoting the mean velocities of the initial primary and secondary waves by V_1 and V_2, we find, from the values of P and S for an arc of 3°, that, for the upper strata of the earth's crust

$$V_1 = 7{\cdot}1 \text{ kms. per sec. and } V_2 = 4{\cdot}0 \text{ kms. per sec.}$$

For greater distances, however, the table shows a continual increase in the values of V_1 and V_2 with the distance, whether the distance be measured along a great circle or along a chord. Measured along a great circle, the values of V_1 for arcual distances of 30°, 60°, 90° and 120° are 8·6, 10·9, 12·6 and 13·6 kms. per sec.; those of V_2 for the same distances are 4·8, 6·0, 6·9 and 7·7 kms. per sec. Measured along a chord, the corresponding values of V_1 would be 8·5, 10·4, 11·4 and 11·7 kms. per sec., and those of V_2 4·7, 5·8, 6·2 and 6·4 kms. per sec. Thus, in both primary and secondary waves, there is a marked increase in the velocity with the distance as measured along either the arc or the chord. We conclude, therefore, that the primary and secondary waves must travel along curved paths and that the velocities of both waves increase with the depth below the surface, from which it follows that the paths on the whole are concave towards the surface.

On the other hand, it is clear, from the straightness of the curve marked L in Fig. 64, that the time varies as the distance measured along the great circle from the epicentre to the station; in other words, that the velocity of the long waves along the surface is constant. From the mean of a large number of observations, O. Klotz finds this velocity to be 230 kms. per min. or 3·8 kms. per sec.†

* J. Milne, *Rep. Brit. Ass.*, 1898, p. 223; O. Klotz, *Bull. Seis. Soc. Amer.*, vol. 7, 1907, pp. 67–71. Turner's table is given in the circulars issued by the Brit. Ass. Seis. Com.; it is reprinted in full in *Proc. Roy. Soc. Edin.*, vol. 39, 1919, p. 198.

† *Bull. Seis. Soc. Amer.*, vol. 7, 1907, pp. 67–71.

152. (ii) **Distance of Epicentre and Time at the Origin.** Though the time-curves of individual earthquakes may vary slightly from Turner's mean time-curves, the value of $S - P$ (that is, the duration of the primary phase) is practically constant for any given arcual distance. If, then, a seismogram provides definite readings for P and S, the corresponding figure for $S - P$ in the table gives in degrees the distance of the epicentre from the station. It is obvious that determinations of the distance made at three widely separated observatories would fix the position of the epicentre (see sect. 164). Further, knowing the distance of the epicentre, we can find from the table the corresponding value of P and thus ascertain the time at which the earthquake occurred at the origin*.

For instance, let P (the epoch of the first primary wave), as determined from a given seismogram, be 12 h. 6 m. 52 s., and let $S - P$ be 674 seconds. From the table, we find that the epicentral distance is 93° (that is, 10352 kms.), and that P is 812 secs. or 13 m. 32 s. Thus, the time at the origin was 12 h. 6 m. 52 s. less 13 m. 32 s., or 11 h. 53 m. 20 s.

153. Nature of the Primary and Secondary Waves. We have seen that the motion of a distant earthquake is divisible into three well-marked phases, of which the primary and secondary waves pursue paths which penetrate the body of the earth, while the principal waves travel across its surface; and this division naturally led to the suggestion (by Oldham in 1899) that the primary waves consist of condensational, and the secondary waves of distortional, vibrations. Later observations, and especially those on the direction of the vibrations, on the whole support this view, though it is clear that condensational and distortional vibrations exist in both phases, the former predominating in the primary waves and the latter in the secondary waves.

It is obvious that, in the outer crust of the earth, neither condensational nor distortional waves could for long maintain their simple character without being split up into waves of both types as they passed from one rock to another (sects. 142–144). The records of seismographs near the epicentre of an earthquake

* This time is usually denoted by the letter O, and the distance of the station from the epicentre by Δ.

cannot therefore be expected to show any separation of condensational from distortional waves; and, indeed, this never takes place within a distance of about 10°, or 700 miles, from the epicentre. Since, at stations beyond this limit, the characteristic triple phases begin to appear on seismograms, it is clear that the waves arriving there must, for part of their journey, have traversed some homogeneous material situated below a comparatively thin layer. On re-entering or entering this layer, the waves which have become simplified in type must again, by numerous reflections and refractions, become complex, although maintaining on the whole their condensational character in the first series and their distortional form in the second.

This explanation accounts for the nature of the two phases and for the greater velocity of the primary waves, but not for the continuity of the vibrations throughout each of the two early phases. Partly, the subsequent vibrations of each series may be due to internal reflection of the initial waves, as shown in sects. 158, 159, but such vibrations would be discontinuous, and the observed continuity and irregularity of the later movements is probably caused by repeated reflections and refractions at the bounding surfaces of the rocks which constitute the outer crust.

154. One other point with regard to these phases remains to be considered. While the weakness of the primary waves seems to be the only obstacle to their registration at great distances, the secondary waves apparently die out at stations more than 110° or 120° from the epicentre. Their place in the series is not unoccupied, but the vibrations which follow the primary waves at such distances are different from the typical secondary waves. Instead of beginning with a strong movement, with a maximum soon attained, followed by a rapid decline in strength, the substituted movement begins gradually, and there is no well-marked maximum, but rather a succession of impulses. Moreover, as G. W. Walker points out (p. 42), the times at which they appear agree with the times at which waves reflected at or near the surface would reach the station (sects. 158, 159). But, whether or no this be the correct explanation of the movements, it is clear that the secondary waves cease to exist as such at distances from the epicentre greater than 110° or 120°.

155. Nature of the Long Waves. While there is general agreement among seismologists as to the origin of the primary and secondary waves, the nature of those which constitute the principal portion of the movement is somewhat uncertain. They are usually identified with surface-waves, known as Rayleigh waves, the existence of which was proved by Lord Rayleigh and H. Lamb. The velocity of the distortional waves in the outer rocks is known to be 4·0 kms. per sec. That of the Rayleigh waves, according to theory, is ·9194 × 4·0, or 3·7 kms. per sec., a result which is in close agreement with Klotz's estimate of 3·8 kms. per sec. for the mean velocity of the initial long waves. Even if this identification be correct, it is still difficult to account for the great duration of this phase of the movement, and especially for that of the end portion or coda, which may last as many hours as the shock does minutes at the epicentre*.

156. Form of Seismic Rays. The determination of the velocities of the primary and secondary waves shows, as already mentioned (sect. 151), that these waves cannot travel along the surface or along chords to the various stations at which they are recorded, but rather along curved paths which, on the whole, are concave towards the surface. The exact form of these paths or seismic rays has been considered by several seismologists, and especially by C. G. Knott, whose latest work is here described.

The only assumptions which Knott requires in this work are: (i) that the earth is a sphere, (ii) that the earthquake originates so near the surface that the depth of the focus, in comparison with the radius of the earth, may be neglected, and (iii) that the elastic properties of the earth's substance depend only on the distance from the centre, and are therefore constant over any sphere concentric with the earth. If these assumptions be correct, every seismic ray which emerges at a given station must lie in the plane through the focus, the station and the centre of the earth. In accordance with the well-known law which governs all elastic waves, every ray must be such that the vibrations

* Rayleigh, *Proc. Math. Soc. Lond.*, vol. 17, 1885, pp. 4–11; H. Lamb, *Phil. Trans.*, vol. 203 A, 1904, pp. 1–42. It has also been suggested that the long waves are those distortional waves which suffer innumerable reflections at the surface (sect. 158) and, as it were, creep round beneath the surface: Knott (2), pp. 256–257.

travelling along it are transmitted in the shortest possible time. In addition, Knott makes use of Turner's table for *P* and *S* given above (sect. 150), and thus any errors that may exist in this table must in so far affect the accuracy of his results.

The forms of seventeen complete seismic rays have been determined, ten for the primary waves and seven for the secondary waves. As both waves are of the same general character, the former only are represented in Fig. 65, which gives a section of a hemisphere through the focus.

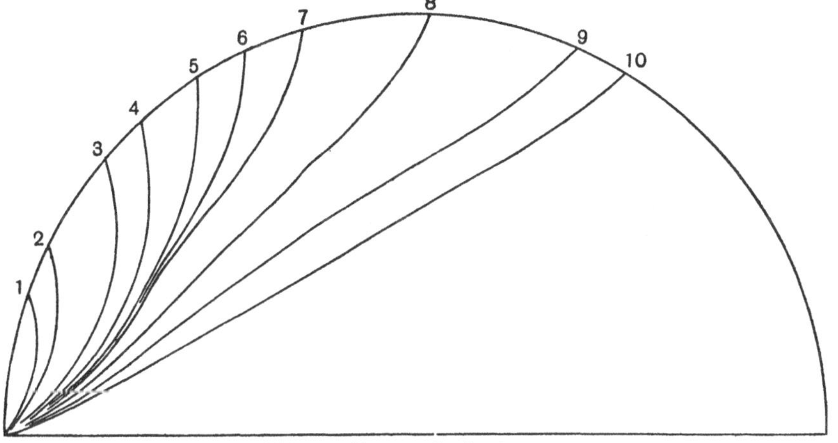

Fig. 65. Forms of seismic rays.

With the exception of the seismic ray directed to the centre of the earth, every ray is notably curved. All those (numbered 1–5) which emerge at distances of less than 60° from the focus are concave towards the surface throughout their whole course. Of those which emerge at greater distances, the central and deeper portion becomes first nearly straight, as in ray 6, and then even slightly convex towards the surface, as in ray 7. This ray emerges at an arcual distance of 73° from the focus. The slight convexity in its form shows that the velocity has begun to decrease as the depth increases. The change takes place at a depth equal to about three-tenths of the earth's radius. The remaining rays (8–10), however, are practically straight throughout a large part of their courses, and this shows that the velocity at depths somewhat greater than the above amount does not vary much with increasing depth.

157. While the forms of the primary and secondary seismic rays are approximately similar, there is, however, a difference to be noticed in the variations of the velocities of the two types of waves in the interior of the earth. These velocities are represented by the curves in Fig. 66, in which depths below the surface in thousands of kilometres are measured along the horizontal axis (the radius of the earth being 6378 kilometres) and

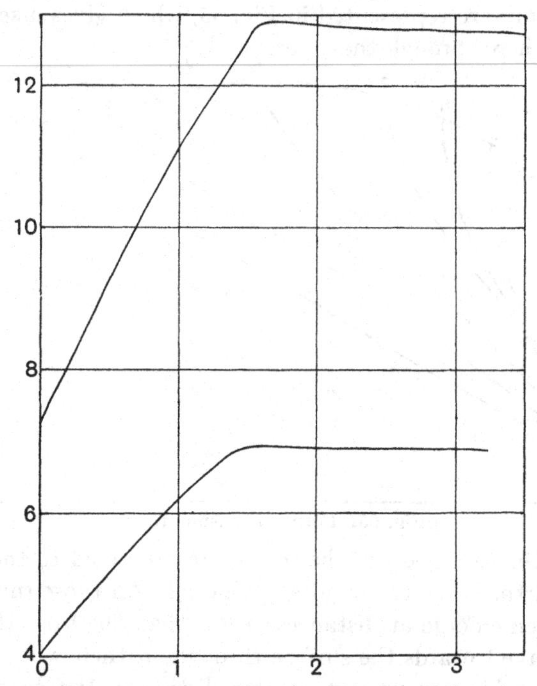

Fig. 66. Diagram representing the variation of the primary and secondary wave-velocities with the depth.

velocities in kms. per sec. in the perpendicular direction*. It will be seen that both curves are practically straight to a depth of about 700 kms.; from 700 to 800 kms., there is a slight bend, after which the curves become again nearly straight to a depth of about 1400 kms. for the secondary wave and about 1600 kms. for the primary wave. At still greater depths, the velocity of

* These velocities, it should be noted, are actual velocities in the direction of the wave-path. The figures in sect. 151 are those for the mean surface velocity, which is of course greater than the actual velocity.

each wave is nearly constant, being 12·8 kms. per sec. for the primary, and 6·85 kms. per sec. for the secondary, wave. It will be seen (sect. 161) that the results of this and previous sections have an important bearing on our knowledge of the nature of the earth's interior*.

158. Internal Reflection of Primary and Secondary Waves. When the primary and secondary waves meet the surface of the earth, they are reflected and refracted (sect. 142). As the second medium is either water or air†, the refracted wave is

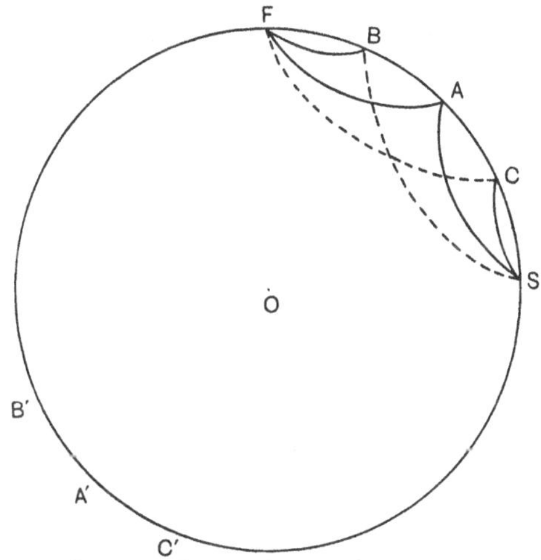

Fig. 67. Diagram illustrating the internal reflection of earthquake-waves at the earth's surface.

condensational and possesses but a small part of the energy of the incident waves; the remainder is distributed between the reflected condensational and distortional waves. These may arrive at a given station after one or more reflections, and, though each wave after reflection will be of diminished energy, it may combine or interfere with waves proceeding by other routes and thus either strengthen or weaken those waves at the observing station.

It is clear that an incident condensational wave may travel with one reflection from the focus F to the station S (Fig. 67)

* Knott (3).

† There may also be reflection and refraction at the lower surface of the outer heterogeneous crust of the earth.

in one of two ways: (i) the condensational wave may be reflected at the mid-point *A* of the arc *FS*, and the reflected condensational wave continue by a similar path to *S*; or (ii) the condensational wave may be reflected at a point *B* between *F* and *A* and the reflected distortional wave may proceed along the deeper course *BS*. Again, an incident distortional wave may be reflected at *A*, and the reflected distortional wave continue to *S*; or the distortional wave may travel along the path *FC* (similar to *BS*) and the reflected condensational wave reach *S* by the shallower path *CS*. Similar reflections may occur at the points *A'*, *B'*, *C'* on the major arc *FS*, and, if the constitution of the interior of the earth permitted their passage, the three waves would ultimately arrive at *S*.

Waves also reach a given station *S* after two, three or more reflections, with or without change of type, and, though greatly weakened by such reflections, may help to reinforce vibrations proceeding by more direct routes. If, however, the angle of incidence be very nearly 90°, we have seen (sect. 142) that the wave may proceed unchanged in type without very material loss of energy. Such a wave would suffer innumerable reflections and practically would creep round the surface in the same way as the sound-waves are known to creep round the dome of St Paul's cathedral*.

159. The question now arises whether such reflected waves can be recognised on seismograms. Let us consider the record of the Bonin Islands earthquake of Nov. 24, 1914, illustrated in Fig. 63, and confine ourselves to single reflections without change of type of the condensational and distortional waves. From the seismogram, it is found (sect. 152) that the arcual distance of the origin is 93° and that the time at the origin was 11 h. 53 m. 20 s. Now, the point of reflection would be at a distance of $46\frac{1}{2}°$ from the focus, for which $P - O$ is $523\frac{1}{2}$ seconds, so that the total time taken by the primary wave to travel from the focus to the point of reflection and on to the station would be $523\frac{1}{2} \times 2$ seconds or 17 m. 27 s. Thus, the time at which the reflected primary wave would reach the station would be 11 h. 53 m. 20 s. + 17 m. 27 s., or 12 h. 10 m. 47 s.

* Rayleigh, *Theory of Sound*, vol. 2, 1878, pp. 115–116; Wiechert, pp. 103–115.

Again, for the epicentral distance of $46\frac{1}{2}°$, the value of $S - O$ is $934\frac{1}{2}$ seconds, so that the time taken by the secondary wave in travelling from the focus to the point of reflection and on to the station would be $934\frac{1}{2} \times 2$ seconds or 31 m. 9 s. Thus, the time at which the reflected secondary wave would reach the station would be 11 h. 53 m. 20 s. + 31 m. 9 s., or 12 h. 24 m. 29 s. It will be seen, from Fig. 63, that the pronounced movements marked *PR* and *SR* occur almost exactly at these times, and it may fairly be concluded that they are the reflected waves for which we are seeking*.

160. Returns of the Long Waves. As the large waves diverge mainly in two dimensions, they retain a large part of their energy at great distances from the epicentre, and it is therefore to be expected that they should cross a given station more than once. In strong earthquakes, they are usually recognised twice, and, in a few earthquakes, they have been recorded three times. In their different transits, they are known as the W_1, W_2 and W_3 waves. If, in the diagram (Fig. 68), the circle represent a section of the earth through the focus F and the station S, it is clear that the W_1 waves are those which travel along the minor arc *FAS*; the W_2 waves those which traverse the major arc *FBS*†, and the W_3 waves those which, after passing S the first time, continue their journey round the world, pass through the focus F, and traverse for the second time the minor arc *FAS*.

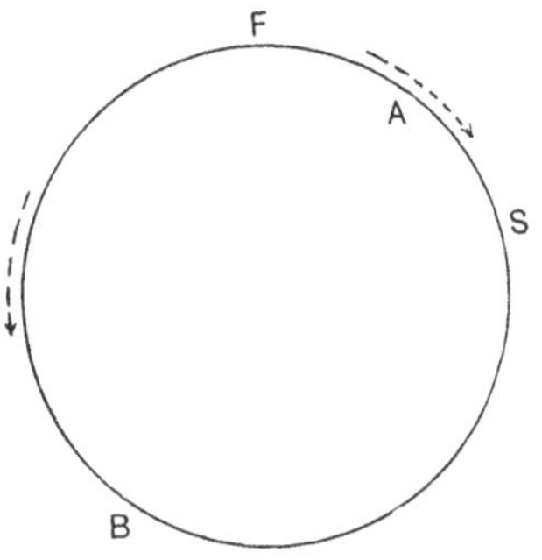

Fig. 68. Diagram illustrating the returns of the long waves.

If t_1, t_2 and t_3 be the observed times of passage of the initial

* It is doubtful whether waves suffering a single reflection at the mid-point of the major arc *FS* (Fig. 67) could be detected. After reflection and so long a course, the primary waves would probably be too weak to influence the record, and, in any case, they would be confused by the long waves among which they would arrive. As the secondary waves are not distinguishable more than 120° from the origin, and as half the major arc to a station at less distance must be greater than 120°, it follows that, after a single reflection at A, they would never reach the station.

† The W_2 waves are clearly shown at W_2 in the record of the Kangra earthquake of 1905 (Fig. 61).

undulations of the W_1, W_2 and W_3 waves, and Δ the length in kilometres of the minor arc *FAS*, the mean velocities of the initial W_2 and W_3 waves are, respectively,

$$\frac{40000 - 2\Delta}{t_2 - t_1} \text{ and } \frac{40000}{t_3 - t_1} \text{ kms. per sec.;}$$

40,000 kms. being the circumference of the earth and $t_3 - t_1$ the time taken by the W_3 waves to travel completely round the world.

The average velocity of the initial W_2 waves is found by Omori to be 3·7 kms. per sec. That of the W_3 waves, as it depends on but a few observations, is somewhat uncertain. From four earthquakes in 1900 and 1902, Omori gives 3 h. 20 m. 46 s. as the mean value of $t_3 - t_1$, and 3·4 kms. per sec. as that of the velocity of the initial W_3 waves. Other estimates of this velocity are 3·53 kms. per sec. for the Messina earthquake of 1908 (Galitzin) and 3·45 kms. per sec. for the Californian earthquake of 1906 (Davison).

It will be noticed that the estimates of the velocity of the W_3 waves are less than those of the W_2 waves. The explanation no doubt is that the W_3 waves are the representatives of the more prominent and somewhat later undulations of the long waves—an explanation supported by Omori's estimates of the mean periods of the third phase (short-period waves) of the principal portion, and of the W_2 and W_3 waves, namely, 20·4, 20·4 and 19·4 seconds*.

Nature of the Earth's Interior

161. As soon as it was realised that the secondary waves consist mainly of distortional waves, which cannot be transmitted by liquids, it was evident that continued seismological observations would extend our knowledge of the nature of the earth's interior. Since then, the chief contributions have been made by Oldham, Wiechert and Knott†. The present section is practically confined to the recent results obtained by Knott.

* F. Omori, *Publ. Eq. Inv. Com.*, No. 13, 1903, pp. 119–124.

† Oldham (2), Wiechert and Knott (3). Oldham's investigations were based on a small number of earthquake-records. The conclusions at which he arrived are: (i) that the interior of the earth, beneath the outer heterogeneous layer (about 30 kms. thick), consists of uniform material that can transmit both condensational and distortional waves, and that this material

(i) The outer heterogeneous crust is clearly thin compared with the radius of the earth. The primary and secondary waves are distinctly separated at a distance of about 10° from the epicentre. The waves which emerge at this distance have penetrated to a depth of about 100 kms. As they must have traversed a considerable course within the homogeneous interior for the sifting-out of the waves to be accomplished, Oldham concludes that the outer crust, consisting of rocks such as we know at the surface, may attain a thickness of 30 kms., or about $\frac{1}{200}$th part of the earth's radius.

(ii) Beneath this outer crust lies a thick and practically homogeneous layer, in which the primary and secondary waves become separated. The seismic rays of both waves are on the whole concave towards the surface, showing that the velocities increase with the depth until that depth is about equal to three-tenths of the earth's radius. After this, the velocities become constant, and then slightly decline for greater depths.

(iii) This elastic solid shell probably extends to a depth of about half the earth's radius, but, since the maximum velocity of the secondary wave is reached some 200 kms. before that of the primary wave, it would seem that, at about the above depth, the rigidity begins to break down.

(iv) The most significant fact is the loss of the secondary waves at a distance of 120° from the epicentre. The distortional waves which emerge at this distance have penetrated to a depth between one-half and six-tenths of the earth's radius. Knott concludes, therefore, that, in the neighbourhood of the latter depth, the elastic solid shell gives place to a non-rigid nucleus of measurable compressibility, capable of transmitting condensational, but not distortional, waves.

Determination of the Epicentre of a Distant Earthquake

162. Of the earthquakes which are recorded at distant stations, the majority originate under the sea, and some of the remainder in countries that are inhabited by uncivilised races. To determine the epicentres of such earthquakes, several methods have been devised. They depend on the approximate constancy with which

undergoes no marked change in physical character to a depth of about six-tenths of the radius; and (ii) that the central four-tenths of the radius are occupied by matter possessing radically different physical properties.

the long waves travel, on the known duration of the preliminary tremors or of the first series only at given distances, and on the direction of the initial movement of the preliminary tremors.

163. (i) **Method depending on Time-Observations** (Milne's first method). This method depends on the time of arrival of the large undulations at four or more places. Let the times at the four places *A*, *B*, *C*, *D* (Fig. 69) be respectively t_1, t_2, t_3, t_4, of which t_1 is the earliest. Let V be the velocity of the long

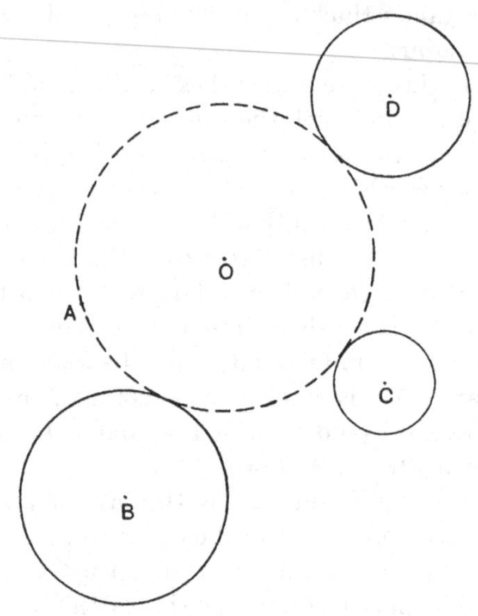

Fig. 69. Diagram illustrating the determination of the epicentre from the times of arrival of the long waves.

waves. When these waves reach the station *A*, they will be at distances $V(t_2 - t_1)$, $V(t_3 - t_1)$, $V(t_4 - t_1)$ from the stations *B*, *C*, *D*. Thus, if circles be described with the stations as centres and the above distances as radii, the first large waves will form a circle which passes through *A* and touches each of the three circles. The centre *O* of this circle is the epicentre of the earthquake.

If the times were known at three stations *A*, *B*, *C*, only, two circles might be drawn through *A* to touch the circles with *B* and *C* as centres. The position of the epicentre might thus be indeterminate.

Milne, who used this method, assumed that the velocity of the large waves was approximately 3 kms. per sec. or 1·6 degrees of arc per min. The circles (represented as on a plane in Fig. 69) were drawn on a slate globe, and the circle with centre O was drawn by trial.

The results obtained by this method are obviously approximate. They depend on the accuracy of absolute time-determinations at the different stations, and these may occasionally err by as much as 1 minute*.

164. (ii) **Method depending on the Duration of the First Preliminary Tremor** (Zeissig's method). If the value of $S - P$ be known for a single station, the epicentre must lie on the circle with the station as centre and the corresponding distance as radius; if the values of $S - P$ be known for two stations, the epicentre must coincide with one of the two points of intersection of two circles; if the values of $S - P$ be known for three stations, the particular point is determined. In practice, of course, the three circles seldom, if ever, intersect in a point. The epicentre is then taken to be the centre of the small triangle formed by the intersections of the three circles.

This method, which is the most frequently used of all, gives results of greater accuracy than the first method, as it depends on measurements of the duration of an interval and not on the absolute times. As a rule, P and S can be determined to within a few seconds; but sometimes the initial tremors are so small that they fail to affect the less sensitive instruments. There may also be some inaccuracy arising from the fact that the tables for $S-P$ give their average values for a large number of earthquakes.

If L denote the time of arrival of the long undulations, similar methods are based on the values of $L - P$ and $L - S$ at three given stations. The first long undulations are, however, less sharply defined than the first movements of P and S, and thus tables depending on the values of L give results that are of inferior value to those obtained by Zeissig's method. The values of $L - S$ may, however, give useful results when the value of P is indeterminate†.

* J. Milne, *Rep. Brit. Ass.*, 1899, p. 33; 1900, p. 79.

† If M denote the time at which the maximum large undulations begin, Milne has used a similar method based on the values of $M - P$ at three stations. O. Klotz has devised a graphical form of Zeissig's method which gives the position of the epicentre with ease and accuracy (*Bull. Seis. Soc. Amer.*, vol. 1, 1911, pp. 143–148).

165. (iii) **Method depending on the Direction and Epicentral Distance** (Galitzin's method). Prince Galitzin has shown that the position of the epicentre may be determined from observations at a single station. From the value of $S - P$ for the station, we know the distance Δ of the epicentre, and thus that the epicentre lies on a circle with the station as centre and Δ as radius. Again, the N.–S. and E.–W. components of the first displacement (which is not complicated by the superposition of other waves) give the direction of the movement, and this is found to coincide very closely with that of the epicentre from the station. The epicentre must thus lie at one or other end of the diameter in this direction, the particular end being determined by the vertical component seismograph, and in the direction of the downward movement. The remarks on the previous method, so far as regards the evaluation of $S - P$, apply to this method.

As an example of this method, Galitzin takes the case of the Monastir earthquake of Feb. 18, 1911. The record of the Galitzin seismograph at Pulkowa (59° 46′ N. lat., 30° 19′ E. long.) gives $\Delta = 2260$ kms. or 20° 19′ of arc, in the direction S. 22° 53′ W. That of the Galitzin seismograph at Eskdalemuir (55° 19′ N. lat., 3° 12′ W. long.) gives $\Delta = 2360$ kms. or 21° 14′ of arc, in the direction S. 55° 58′ E. The former record locates the epicentre in 40·5° N. lat., 20·1° E. long., the latter in 40·6° N. lat., 20·3° E. long.*

166. (iv) **Method depending on the Direction only** (Galitzin and Walker's method). The last method leads naturally to the simplest of all methods, that which determines the epicentre by the intersection of the lines of direction at two stations. It possesses several advantages over other methods, in being independent of all estimates of the time, of the exact determination of the beginning of the second phase, and of the empirical tables for $S - P$. Lastly, although only two stations are used, there is no ambiguity as to the position of the epicentre.

Applied to the Monastir earthquake of Feb. 18, 1911, the azimuths at Pulkowa and Eskdalemuir assign 40·3° N. lat., 20·4° E. long. as the position of the epicentre, a result which agrees very closely with those obtained for the same places by Galitzin's method†.

* B. Galitzin, *Vorlesungen über Seismometrie* (1914), pp. 401–427; *Compt. Rend. Acad. Sci. Paris*, vol. 150, 1910, pp. 642–645, 816–819.

† *Nature*, vol. 90, 1912, p. 3.

CHAPTER X

GEOGRAPHICAL DISTRIBUTION OF EARTHQUAKES

167. De Montessus de Ballore has proposed a rough classification of countries depending on the number and strength of the earthquakes which visit them. He divides them into: (i) *seismic* countries, in which earthquakes are frequent and sometimes disastrous; (ii) *peneseismic* countries, in which earthquakes are severe, but fall short of destructive power; and (iii) *aseismic* countries, in which earthquakes are feeble or rare or even completely unknown. Japan may be taken as a type of the first class, Switzerland of the second, and Russia or Brazil of the third, though, even in Brazil, 50 earthquakes have been recorded since the year 1560*.

Some conception of the varying seismicity of different countries may be obtained from the following figures. In Great Britain, 250 earthquakes originated during the 21 years 1889–1909. On the other hand, 8331 earthquakes were recorded in Japan during the eight years 1885–1892, and 3187 in Greece during the six years 1893–1898. Thus, taking area into account, for every earthquake felt in Great Britain, there were 50 in Japan and 158 in Greece.

168. In constructing a seismic map of a country, either earthquake-frequency alone is represented, or frequency in combination with intensity. De Montessus de Ballore remarks, however, that in some cases frequency and intensity are closely related. In Japan, for instance, he finds that, when the average disturbed area during a given interval increases or decreases, so also does the number of earthquakes in the same interval. The relation does not by any means hold uniformly, for, as at Lisbon in 1755 and Charleston in 1886, a disastrous shock may occur at a place in which earthquakes are far from numerous;

* The principal work on the geographical distribution of earthquakes is F. de Montessus de Ballore's *Géographie Séismique*, 1906, pp. 1–475.

but it is probably general enough to suggest that regions in which earthquakes are frequent are also those in which they are severe.

CONSTRUCTION OF SEISMIC MAPS

169. The object of seismic maps is to depict for given periods the regions of the world or of a country which are specially subject to earthquakes. Such maps may be divided into three classes according as (i) disturbed areas, (ii) meizoseismal areas, or (iii) epicentres, are represented.

170. (i) **Mapping of Disturbed Areas.** Mallet's seismic map of the world is founded on his great catalogue of recorded earthquakes from B.C. 1606 to A.D. 1842; and the method which he uses depends partly on the area disturbed and partly on the intensity. He divides the earthquakes catalogued into three classes: (i) great earthquakes, like the Lisbon earthquake of 1755; (ii) moderate earthquakes, resulting in some damage to buildings but in little loss of life; (iii) minor earthquakes. The disturbed areas of these classes were coloured with different tints, the depths of which were as the numbers 9, 3, 1. When the disturbed areas were unknown, Mallet assumed that the radii of the areas in the different classes were respectively 540, 180 and 60 geographical miles. The mapping of disturbed areas obviously leads to a masking or smoothing away of details. Mallet's map, however, led to the discovery of the principal law which governs the distribution of earthquakes*.

171. (ii) **Mapping of Meizoseismal Areas.** In his studies on the seismic regions of Italy, M. Baratta used a more detailed method: (i) the meizoseismal areas are mapped; and (ii) the intensity, as well as the frequency, of the earthquakes is represented by means of an arbitrary scale of nine degrees, ranging from one very strong shock or several rather strong shocks to two very disastrous earthquakes. Baratta's map of southern Calabria, in which the principal meizoseismal areas are represented, is shown in Fig. 73†.

172. (iii) **Mapping of Epicentres.** De Montessus de Ballore's seismic maps are based on his great catalogue of nearly 160,000

* *Rep. Brit. Ass.*, 1858, pp. 57–72 and plate 12.

† *Boll. Soc. Geog. Ital.*, 1905, p. 1080.

earthquakes. As a general rule, earthquakes disturb an area less than that of a circle 25 miles in diameter, and their epicentres must lie nearer than $12\frac{1}{2}$ miles to the place shaken. On the maps, each place reputed as an epicentre is indicated by a small circular mark, the diameter of which represents the number of earthquakes felt there*.

173. Milne's seismic map of Japan is founded on his catalogue of 8331 earthquakes recorded between the years 1885 and 1892. The map of the country is divided by N.–S. and E.–W. lines into a network of more than 2000 rectangles, the sides of which are respectively one-sixth of a degree of latitude and longitude in length. On his map of the country, Milne represents the epicentres in each rectangle by dots which are distributed uniformly over the rectangles in which they occur. He divides the whole country into 15 districts, but in two of them the epicentres are so numerous that they spread beyond the natural boundaries of the districts, and, in one case, occupy an area of more than two degrees from north to south and nearly two degrees from east to west†.

174. Davison's map of Japan is also founded on Milne's catalogue, its object being to represent the details of distribution with greater precision. It consists in drawing curves through the centres of all rectangles which contain the same number of epicentres, or through points which divide the lines joining the centres of consecutive rectangles in the proper ratio. For instance, in drawing the curve marked 10, if there were nine epicentres in one rectangle and thirteen in the next, the line joining the centres of the rectangles would be divided into four equal parts, and the curve would be drawn through the point of division nearest to the centre of the rectangle containing nine epicentres. The seismic map of Japan drawn in this manner is

* In O'Reilly's seismic map of Great Britain (*Trans. Roy. Irish Acad.*, vol. 28, 1884, plate 40), a somewhat similar method is employed, but every place at which an earthquake was felt is indicated. O'Reilly's method thus differs from Mallet's chiefly in the representation of disturbed areas by a series of dots instead of washes of colour. In his earlier memoirs, de Montessus de Ballore represented the seismicity of a region by lines at right angles to one another forming squares, the sides of which were proportional to $\sqrt{(pA/n)}$, where n is the number of earthquakes recorded in p years in the region of area A sq. kms.

† *Seis. Journ.*, vol. 4, 1895, pp. i–xxi.

reproduced in Fig. 70. The maps for certain smaller districts are given in Figs. 83, 86 and 87*.

SEISMIC MAPS OF THE WORLD

175. The construction of Mallet's great map is described in sect. 170. The general conclusions which he draws from his map

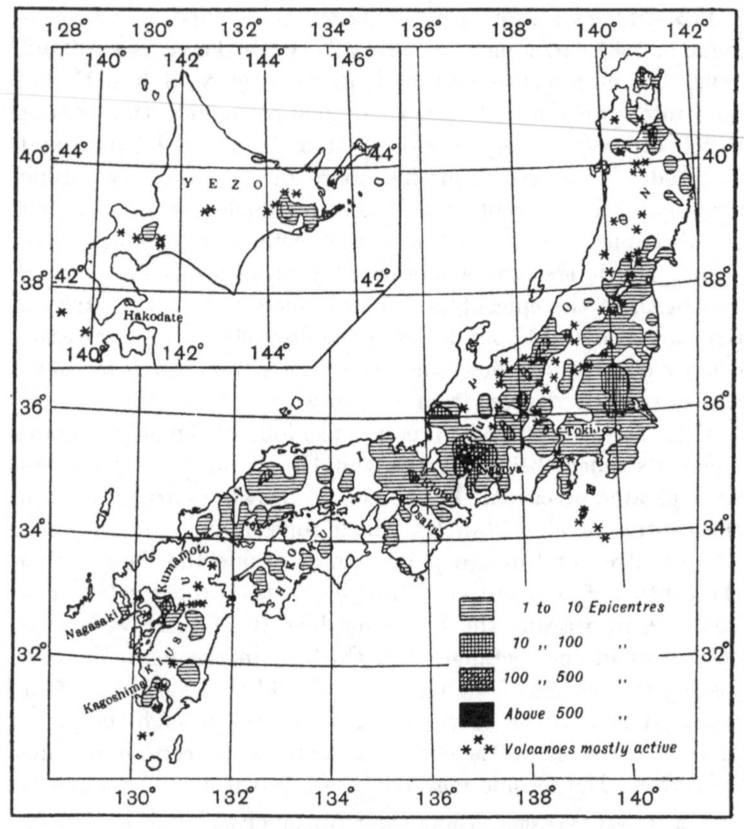

Fig. 70. Seismic map of Japan (1885–1892).

are the following: (i) The normal type of distribution is that of bands of variable but great breadth, extending from 5° to 15°. (ii) These bands generally follow the lines of elevation which mark and divide the great oceanic or terr-oceanic basins of the earth's surface. (iii) In so far as these are frequently the lines

* *Geogr. Journ.*, vol. 10, 1897, pp. 530–535.

of mountain-chains, and these again of volcanic vents, so the seismic bands are found to follow them likewise. (iv) The areas of minimum or no disturbance are the central portions of great oceanic or terr-oceanic basins and the greater islands existing in shallow seas*.

In these conclusions, Mallet foreshadows the law, which is manifested more clearly in detailed maps, that earthquakes are numerous where the gradient of the surface is considerable and rare where the gradient is small.

176. De Montessus de Ballore's seismic map of the world is based on his great catalogue (1903), which contains the entries of nearly 160,000 earthquakes. These earthquakes are of every degree of intensity, the great majority of them being very slight; and, as all known earthquakes are included, it is obvious that this map represents the distribution of seismic energy in space during the whole of historic time.

The principal law which de Montessus de Ballore deduces from his map is the following: The earth's crust trembles almost only along two narrow zones which lie along two great circles of the earth, called by him the Mediterranean or Alpine-Caucasian-Himalayan circle, and the circum-Pacific or Ando-Japanese-Malayan circle.

It is not to be inferred that the instability of these bands is uniform throughout, or even that instability characterises their entire course. Here and there, the bands are interrupted by peneseismic, and even by aseismic, regions. All that is asserted in the law is that the chief seismic districts are situated along these bands, and, that this is the case, will be evident from the following table, in which the second column contains the number of known earthquakes (up to 1906) and the last column the percentage of the total number of earthquakes:

	No. of earthquakes	Per cent.
Mediterranean circle	90126	52·57
Circum-Pacific circle	66026	38·51
European continental area	8939	5·21
Extra-European continental areas	6343	3·71

* *Rep. Brit. Ass.*, 1858, pp. 71–72.

Thus, the seismic regions lying along the two great circles include 91 per cent. of all known earthquakes, while the continental areas (notwithstanding their much greater extent) contain only 9 per cent.

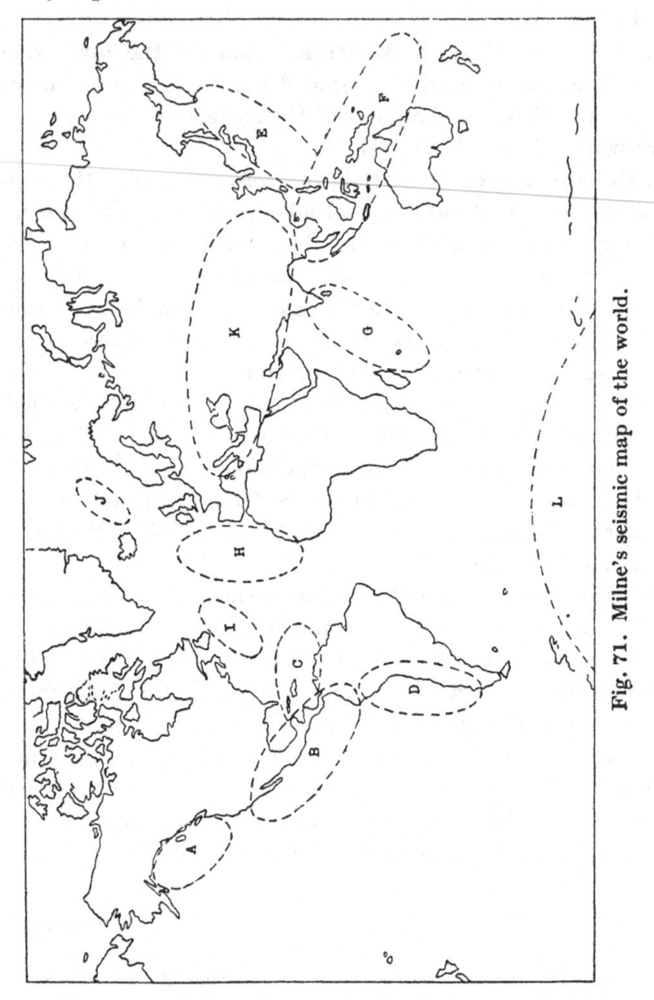

Fig. 71. Milne's seismic map of the world.

A distribution so remarkable must be capable of some explanation, and this de Montessus de Ballore finds to be that the zones enclosing the seismic regions coincide exactly with the geosynclinals of the secondary epoch, as outlined by Haug.

Thus, the unstable bands of the earth are those in which sediments of great thickness have been intensely folded, dislocated and elevated in Tertiary times when the principal existing mountain-chains were formed; while the stable portions of the earth are those connected with the tabular architecture of the great continental regions*.

177. Very different in the materials employed, but not less interesting, are Milne's seismic maps of the world, which he issued annually in the reports of the Seismological Committee of the British Association. In dealing only with earthquakes felt over an area not less than that of Europe and Asia combined, Milne has greatly simplified the problem of seismic distribution. His maps, one of which is reproduced in Fig. 71, show the regions of the earth in which the epicentres of the ten years 1899–1908 are situated†. Their only defect is not one of principle, but is simply due to the brevity of the period considered; for, in the history of terrestrial change, ten years is a length of time almost negligible. As will be seen in a later section (sect. 183), seismic activity on a great scale is subject to considerable migration. Except for minor shocks, a region once struck by a violent earthquake tends to remain quiescent for a prolonged interval. In this respect, Milne's maps are less instructive than de Montessus de Ballore's. On the other hand, in their completeness, they possess a compensating merit, for no great earthquake, wherever it may occur, can now escape detection and registration.

In the map given in Fig. 71, twelve districts are represented, roughly bounded by oval curves‡. It will be seen that four of them (*G*, *H*, *I* and *J*) are entirely oceanic, one of them (*K*) terrestrial, while six (*A*, *B*, *C*, *D*, *E*, *F*) are partly oceanic and partly terrestrial. As the district *C* (including the West Indies and the Caribbean Sea) belongs technically to the Pacific, it will be seen that these six districts cling to the east and west margins of that ocean.

* De Montessus de Ballore, pp. 23–26.

† The method of determining the position of the epicentre is explained in sect. 163.

‡ This map is reproduced from the *Report of the Seismological Committee* for 1909. In the latest reports, the maps are more detailed and less suited for the purpose of this chapter. They should, however, be consulted.

The numbers of earthquakes originating during the years 1899–1908 in the different districts are as follows: *A* 40, *B* 55, *C* 30, *D* 28, *E* 133, *F* 175, *G* 26, *H* 35, *I* 5, *J* 5, and *K* 141. Thus, the four oceanic regions contain 11 per cent. of the total number of earthquakes, the great terrestrial region (*K*) 21 per cent., and the six regions (*A–F*) bordering the Pacific Ocean 68 per cent. Thus, roughly, of every ten great earthquakes, seven occur near the margins of the Pacific, two in the great land-region, and one in the oceanic regions.

At the present time, the great unstable region of the earth lies along the west margin of the Pacific, including Japan, the Philippine Islands and the Malay Archipelago. Of the earthquakes in the six Pacific regions, the three along the east margin contain 33 per cent. of the total number, while those along the west margin contain 67 per cent. Thus, approximately, for every two great earthquakes felt along the east margin, there are five along the western side.

Laws of Seismic Distribution

178. Connexion with the Gradient. The most important law of seismic distribution is that earthquakes are as a rule strongest and most frequent in those portions of the earth in which the average slope of the ground is greatest.

On a large scale, this fact is evident from the maps of the whole world described in sects. 175–177. The bands of seismic activity depicted by Mallet generally follow the lines of elevation which divide the great basins of the earth's surface and shun the central portions of those basins. Again, the margins of the Atlantic slope gently towards the central basin, except in the Gulf of Mexico and the Caribbean Sea; those of the Pacific, and especially along its western border, dip steeply towards the great known depths of ocean. Now, the Atlantic earthquakes amount to 7 per cent. of those registered by Milne during the years 1899–1908. In contrast with these figures, the Pacific earthquakes number 71 per cent. of the whole, those on the west side 48 per cent. and those on the east side 23 per cent.

Again, the great terrestrial region (*K*) includes four sub-regions, namely, the Alpine, the Balkan, the Caucasian and the

Himalayan. It is in the latter, which includes 63 per cent. of the earthquakes belonging to the whole region, that the most pronounced foldings occur, and that, in distances of 100 miles, gradients of 1 in 44 are to be found.

179. The same law holds true for smaller areas than those considered in the last section. De Montessus de Ballore, who included in his survey earthquakes of all degrees of strength, states the law in the following precise manner: "in a general way, we may say that, of two contiguous regions—for example, the two slopes of a valley, the two flanks of a mountain-chain, plains and neighbouring heights, etc.—the more unstable is that which presents the greater average slope." In the same way, Milne, in his study of the Japanese earthquakes of 1885–1892, concludes that "where there is the greatest bending, it is there that sudden yielding is the most frequent." The epicentre of the Sanriku earthquake of June 15, 1896, for instance, was at a depth of 4000 fathoms, exactly at the bottom of the western slope of the well-known Tuscarora Deep (Fig. 72).

The earthquakes of the Japanese Empire are worthy of study in greater detail. These islands are arranged in the form of a festoon with its convexity facing the Pacific Ocean, and, as in similar groups of islands and in mountain-chains of the same form, the convex side of the festoon slopes more steeply than the concave side. As already indicated, the earthquakes of this country follow a law which is general in such cases; they are more numerous and more violent on the steeply sloping convex side than on the other.

The dotted lines in Fig. 72 represent contour-lines of the sea-bed (for each thousand metres) on both sides of the country. The Japan Sea is shallow, its greatest depth being only 3000 metres (1646 fathoms). The gradient of the sea-bed varies from 1 in 67 to 1 in 110; the average gradient to a depth of 1000 metres for the coast of Hokkaido is 1 in 220. On the other hand, the Pacific Ocean is very deep. The extraordinary basin, called the Tuscarora Deep, reaches a depth of 8000 metres (4376 fathoms) at distances of 110 to 240 miles from the coasts. The gradient is unusually steep, being 1 in 27 off the coast of Nemuro, 1 in 30 off the north-east coast of the main island, and 1 in 16 (to a depth of 3000 metres) off the south-east coast of Kazusa and Awa.

The origins of 221 destructive Japanese earthquakes from the fifth century to the present time have been investigated by Omori. Of the total number, 114 originated inland, 47 under the Pacific Ocean, 17 under the Sea of Japan, 2 under inland seas, while the epicentres of 41 earthquakes are unknown. Ten of these earthquakes were very violent, and, while three of them

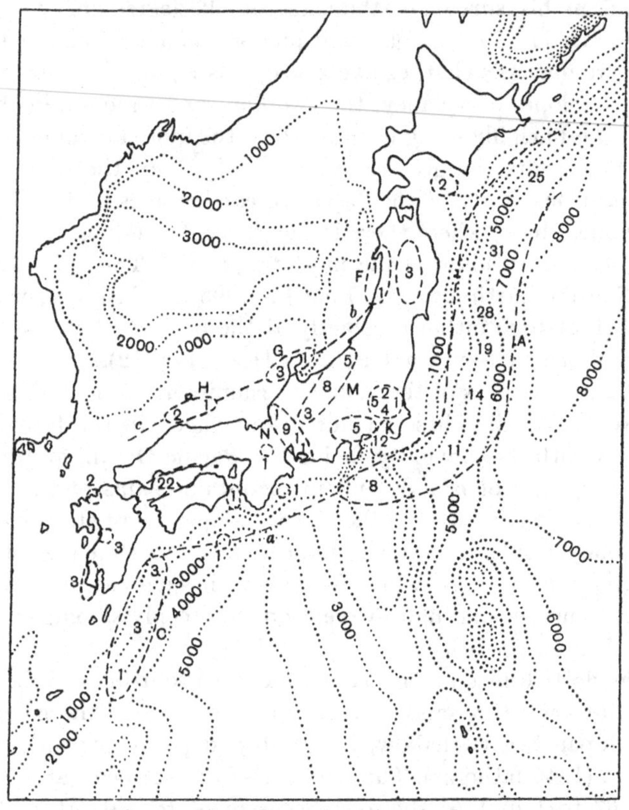

Fig. 72. Map of seismic regions in Japan.

occurred in central Japan, seven originated off the south-east coast, each of the latter being accompanied by seismic sea-waves. Again, on the Pacific coasts, there were 23 great sea-waves during the period mentioned, while on the Japan Sea coast there were only five small sea-waves.

Similar results follow from Omori's examination of recent strong Japanese earthquakes. From 1885, when the systematic observation of earthquakes was begun, to 1905 (that is, in 21 years), 257 earthquakes originated in or around Japan, some of which were destructive or semi-destructive, while the rest were strong or moderate shocks, each having disturbed a land-area of more than about 25,000 sq. miles.

The principal regions into which the epicentres of these strong shocks are grouped are represented by the broken lines in Fig. 72, which is reproduced from Omori's map. During the 21 years referred to, 138 earthquakes originated in the region *A*, 7 in the region *C*, 2 in *F*, 4 in *G*, 3 in *H*, 12 in *K*, 16 in *M*, 12 in *N* and 23 in *P*.

The most active region at present is thus that marked *A*, stretching off the east coasts of Hokkaido and Main Island, the number of earthquakes which occurred in it being rather more than half the total number in Japan. The zone *a*, indicated by a broken line, represents approximately the epicentre of the great earthquake of Oct. 28, 1707 (the greatest of all Japanese earthquakes), and the two great shocks of Dec. 23 and 24, 1854. The zone *a* evidently connects the regions *A* and *C*, the whole forming the principal sub-oceanic earthquake-region of Japan, and called by Ōmori the *External Seismic Zone*. The transference of activity, from the zone *a* in former times to the region *A* in the present day, is worthy of notice.

On the Japan Sea side, there are three regions, *F*, *G* and *H*, including between them not more than nine earthquakes. The most important region is *F*, which in past times has produced some violent shocks. The zones, *b* and *c*, represent approximately the positions of the epicentres of the great earthquakes of Dec. 7, 1833, and March 14, 1872. These two zones, with the regions *F*, *G* and *H*, thus form a continuous region bordering the concave side of the Japanese islands, and called by Omori the *Inner Seismic Zone*.

During the 21 years, 1885–1905, there were thus nine earthquakes in the inner, and 145 earthquakes in the outer, seismic zone. In other words, for every strong earthquake originating on the concave Japan Sea side of the islands, there were 16 on the convex Pacific side*.

* De Montessus de Ballore, pp. 18–19; J. Milne, *Seis. Journ.*, vol. 4, 1895, pp. xv–xvi, and *Rep. Brit. Ass.*, 1897, pp. 29–30; F. Omori, *Bull. Eq. Inv. Com.*, vol. 1, 1907, pp. 114–123.

180. The earthquakes of the Japanese Empire have been considered in some detail, for they have been studied more carefully than those of any other country. In other groups of islands arranged in the festoon form, such as Sumatra and Java and the Aleutian Islands, or in mountain-chains like the Himalayas and those of Alaska, the same law of distribution prevails. The steep convex side is visited by more frequent and more violent earthquakes than the gently-sloping concave side.

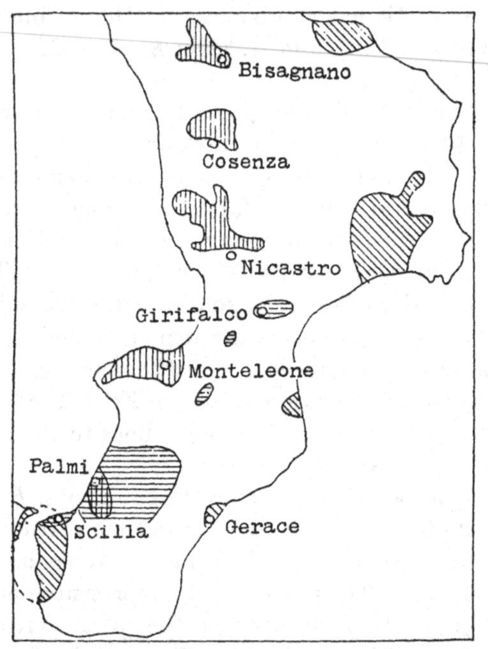

Fig. 73. Distribution of earthquake-zones in Calabria.

181. Earthquakes and Secular Changes of Elevation. A possible relation, on which further information is required, is one pointed out by Milne regarding the occurrence in Japan of districts which are known to be undergoing a slow process of elevation. Of the 15 seismic districts into which he divides the whole country, ten show evidences of very recent elevation, and in five of them earthquakes are extremely frequent. There are, however, two exceptions to this rule, which he points out. One seismic district is, so far as known, free from any movement, either of depression or elevation; while another section of the coast, which is known to be rising, is comparatively free from earthquakes*.

MIGRATIONS OF SEISMIC ACTIVITY

182. Small Migrations of the Epicentre. Detailed mapping of successive earthquakes in any seismic region shows that the epicentres are subject to continual migration. Two examples of this migration are given in other chapters, the after-shocks of the Inverness earthquake of 1901 being considered in sect. 240, and those of the Mino-Owari earthquake of 1891 in sect. 215.

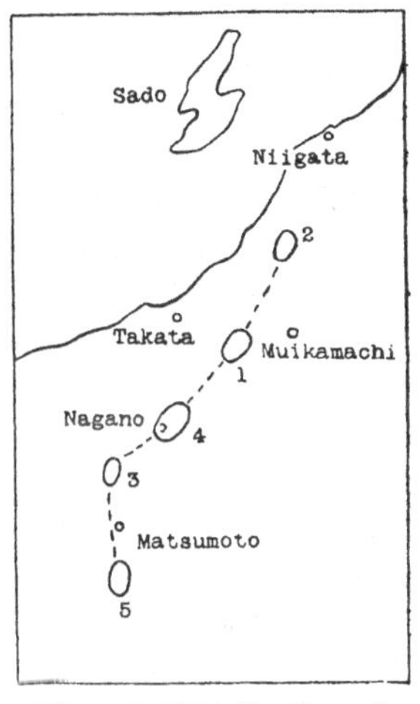

Fig. 74. Distribution of earthquakes in central Japan.

The migration of earthquake-centres in another district, that of southern Calabria, is represented in Fig. 73. The shaded areas are those most strongly shaken in different earthquakes, and therefore must closely surround the corresponding epicentres. The areas shaded with horizontal lines belong to the remarkable series of Calabrian earthquakes in 1783 and the following years. Those shaded with vertical lines relate to the earthquakes of 1905, while those shaded obliquely are the central areas of other earthquakes.

In 1783, the first great earthquake occurred on Feb. 5 in the Palmi zone, the second a few hours afterwards in the Scilla zone. Two days later, the third great earthquake of the series took place in the Monteleone zone, followed in two hours by the fourth in the Messina and Scilla zones. On Mar. 5, the fifth great earthquake occurred in the Monteleone zone, and, on Mar. 28, a sixth, almost, if not quite, as strong as the first of the series, in the Girifalco zone. Sometimes, the different areas are shaken simultaneously, or perhaps in very rapid succession. In the great earthquake of 1905, this was the case in five zones, namely, those of Palmi, Monteleone, Nicastro, Cosenza and

* *Seis. Journ.*, vol. 4, 1895, p. xvi.

Bisagnano. At other times, single areas only are affected, such as that of Nicastro in 1638, Montcleone in 1659, Bisagnano in 1836, Cosenza in 1854, Palmi in 1894, and Messina in 1908.

The district represented in Fig. 74 is the Shinano-gawa earthquake-zone in Japan (*M*, in Fig. 72). The small ovals indicate the approximate positions of the epicentres, the numerals attached to them showing the order of their occurrence. All were severe local shocks, strong enough to produce cracks in the ground and to cause some injury to buildings. The first earthquake of the series occurred on July 23, 1886, and the successive shocks on July 22, 1887, Jan. 7, 1890, Jan. 17, 1897 and Jan. 22, 1899*.

183. Large Migrations of the Epicentre. The transference of seismic activity is equally characteristic of larger areas. For instance, in the Iberian peninsula, the western portion was chiefly affected in the eighteenth century, and the southern districts in the nineteenth. In the former century, Portugal was frequently shaken, the great Lisbon earthquake tending to close the series in 1755. In the next century, Portugal was almost unshaken, while destructive earthquakes occurred in Almeria in 1804, 1860 and 1863, in Murcia in 1828–29 and 1864, and in Andalusia in 1884.

On a still larger scale is the migration of seismic activity along the west coast of the American continent from 1899 to 1912. On Sep. 4 and 11, 1899, and Oct. 9, 1900, there were three great earthquakes in Alaska; on Jan. 20, 1900, and Apr. 19 and Sep. 23, 1902, there were destructive shocks in Mexico, Guatemala and other parts of Central America; on Jan. 31, 1906, a seventh in Panama and along the west coast of Colombia and Ecuador, followed on Apr. 18 by the Californian earthquake, and on Aug. 17 of the same year, by the great earthquake in Valparaiso. On Apr. 15, 1907, and Nov. 19, 1912, disastrous earthquakes took place in Mexico. By the epicentres of these eleven earthquakes, the western coast of America may almost be said to have been outlined from the extreme north-west to as far south as Chili†.

* M. Baratta, *Sopra le zone sismologicamente pericolose delle Calabrie*, Voghera (no date); F. Omori, *Bull. Eq. Inv. Com.*, vol. 1, 1907, pp. 138–141.

† *Bull. Eq. Inv. Com.*, vol. 1, 1907, pp. 21–23.

Distribution of Submarine Earthquakes

184. Submarine earthquakes belong to three classes: (i) those which originate so near land that they are felt along the adjoining coasts; (ii) those which are felt on passing ships; and (iii) those which are strong enough to be registered at some or all seismological stations. The distribution of earthquakes in the third group is considered in sect. 177, from which it is clear that a large number of great earthquakes are certainly of submarine origin. How frequent are earthquakes of the first class is evident from the observations described above (sect. 179). In a seismic country like Japan, it would seem that at least half the earthquakes originate beneath the ocean.

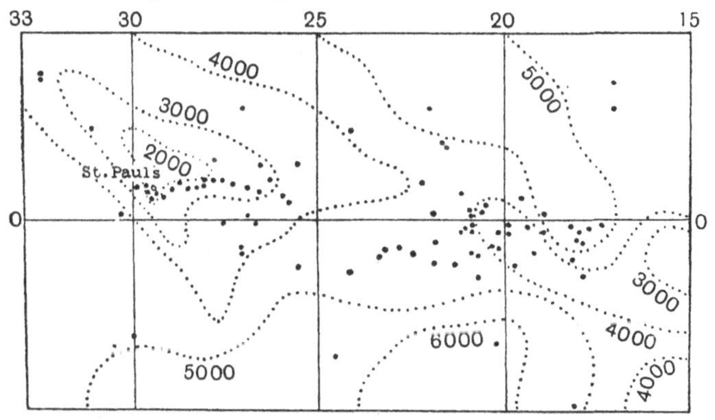

Fig. 75. Principal seismic regions of the equatorial Atlantic.

Our knowledge of the second class of earthquakes is the least complete of the three, and is chiefly derived from the logs of ship-captains, extracts from which have been collected and analysed by E. Rudolph. The uniform experience in such seaquakes is of a sharp vertical motion, as if the ship were grinding over a reef of rocks, the shock being accompanied by a low rumbling sound (sect. 142). In order to estimate the intensity of the seaquakes, Rudolph has devised an arbitrary scale, the corresponding degrees of the Rossi-Forel scale and the number of seaquakes belonging to each degree of the Rudolph scale being given in the following table:

Rudolph scale	1	2	3	4	5	6	7	8	9	10
Rossi-Forel scale	3			4		5, 6	7	8	9	10
No. of seaquakes	8	6	15	14	65	52	14	22	19	7

That the records are incomplete is clear from the greater numbers for recent years, the larger number felt in the Atlantic (127) than in the Pacific (93), and the rarity of the cases in which a seaquake is felt in more than one ship. It can only be along the most frequented ocean-routes that the majority of disturbances are noticed. Of 333 seaquakes recorded by Rudolph, it appears that 52 were felt in the North Atlantic, 65 in the equatorial Atlantic, and 10 in the South Atlantic, 34 in the Gulf of Mexico, etc., 31 in the Mediterranean, 28 in the Indian Ocean, and 20 in the East Indian archipelago, 69 along the east side, and 24 along the west side, of the Pacific. The two principal seismic regions of the equatorial Atlantic are represented in Fig. 75, the dots showing the positions of the ships at the times the seaquakes were felt.

In their distribution, seaquakes closely resemble earthquakes. There is the same clustering in some regions and absence from others, the same independence of volcanoes, and they occur at all depths below the sea-level. It is, however, at the foot of the steep slopes that border continents and islands that the greatest changes of contour occur, resulting in the frequent rupture of telegraph-cables. On the nearly level deep-lying ocean-floors, such cables remain uninjured for many years*.

* J. Milne, *Geogr. Journ.*, vol. 10, 1897, pp. 129–146, 259–285; E. Rudolph, *Beitr. zur Geoph.*, vol. 1, 1887, pp. 133–365; vol. 2, 1895, pp. 537–666; vol. 3, 1898, pp. 273–336. Many cable ruptures, as Milne shows, are due to submarine landslips.

CHAPTER XI

FREQUENCY AND PERIODICITY OF EARTHQUAKES

FREQUENCY OF EARTHQUAKES

185. Annual Frequency of Earthquakes. Any estimate that may be formed of the annual frequency of earthquakes depends on the definition chosen for a unit earthquake, on the lower limit accepted for the perceptible intensity, and on the completeness of the available catalogues. The older estimates (such as those of Perrey and Mallet) are inevitably defective, owing to the absence of information from some countries, to the general inattention as regards all but disastrous shocks in others, and to lack of instruments capable of recording disturbances that are imperceptible to man.

Thus, Perrey found the average number of shocks felt in Europe and the adjoining portions of Asia and Africa during the ten years 1833–1842 to be 33 a year. Mallet, in his catalogue of recorded earthquakes from the earliest times to 1842 tabulated 6831 shocks, of which 216 were "great," that is, strong enough to reduce whole towns to ruins. During the first half of the nineteenth century, the number of earthquakes recorded by him was 3240, of which 53 were "great." Thus, during this half-century, the average annual number of earthquakes was 65, including about one "great" shock. De Montessus de Ballore in 1900 raised the probable annual number to 3830, and Milne a few years later estimated the total annual number of earthquakes and earth-tremors at not less than 30,000.

186. De Montessus de Ballore's method of estimating the annual frequency of earthquakes is worthy of notice. He divides seismic records into three classes—historical, seismological and seismographic—according as they are based on works in which only important disturbances are noted, on the careful inquiries of one or more observers without instrumental aid, or on the records of several or many seismographs established throughout a country. In 93 regions, he was able to compare the numbers of earthquakes registered in two or even three of the different classes of records. He found that the annual frequencies obtained from the second and first records were as 4·26 to 1, from the third and first as 26·59 to 1, and from the third and second as 6·44 to 1*. Making use of these ratios, he calculates for each district the equivalent number of sensible earthquakes (such as would be given in records of the second class) and finds the annual number to be 3830. That this is by no means excessive is evident from the fact that the catalogue for 1907 issued by the Central Bureau of the International Seismological Association contains the records of 4383 earthquakes. Thus, the average annual number of earthquakes sensible to human beings is probably about 4000†.

187. Frequency in relation to Intensity. Of the 4000 earthquakes felt every year, by far the larger number are feeble shocks. In Japan, the average annual number recorded during the ten years 1885–1894 was 1270 and during the six years 1902–1907 it was 1605; yet, according to Omori, the number of destructive earthquakes which visited the country between A.D. 416 and 1898 was only 222. From 1601 to 1898, the number was 108, or one on an average every $2\frac{3}{4}$ years. Milne has recently compiled an important catalogue of destructive earthquakes during the Christian era. During the 1893 years, from A.D. 7 to 1899, the total number of entries is 4151. In the last 99 years, the number is 423, or between four and five a year. The number of world-shaking or semi-world-shaking earthquakes registered by the Seismological Committee of the British Association is 673 during the ten years 1899–1908, or an average of 67 a year. Taking the latter number, it would thus appear that one out of every 60 or 70 sensible earthquakes is capable of being instrumentally registered over a hemisphere or the entire globe.

Turning to particular countries, we find that, of the earth-

* These three ratios were obtained independently. A test of their accuracy is furnished by the fact that the ratio for the second and first classes given by the ratio 26·59 : 6·44 is 4·18 to 1.

† J. Milne, *Proc. Roy. Soc.*, series A, vol. 77, 1906, p. 367; F. de Montessus de Ballore, *Beitr. zur Geoph.*, vol. 4, 1900, pp. 331–382; E. Scheu and R. Lais, *Catalogue Régional des Tremblements de Terre ressentis pendant l'année* 1907 (Strasbourg, 1912).

quakes recorded in Japan from 1885 to 1892, 81 per cent. disturbed areas less than 1000 sq. miles, 15 per cent. between 1000 and 10,000 sq. miles, and 4 per cent. more than 10,000 sq. miles. Again, dividing earthquakes into three classes—slight, strong and destructive—according as the intensities (Rossi-Forel scale) were 3–6, 7–8 and 9–10, 76 per cent. of the Japanese earthquakes entered in Sekiya's catalogue (A.D. 416–1864) were slight, 13½ per cent. strong, and 10½ per cent. destructive. Of the Italian earthquakes from 1873 to 1885, 79 per cent. were slight, 12½ per cent. strong, and 8½ per cent. destructive. Thus, in a seismic country in which earthquakes are carefully studied, of every 20 earthquakes, 15 are slight, 3 strong and 2 destructive*.

188. Clustering of Great Earthquakes. Every catalogue of destructive earthquakes shows that, while such earthquakes sometimes occur singly, they tend nevertheless to cluster in groups. Omori, for instance, finds that the 154 destructive Japanese earthquakes which have occurred since the beginning of the fourteenth century are divisible into 41 groups of varying duration and number. The greatest number of earthquakes in any group was 12, and the longest duration of a group 15 years. The intervals between successive epochs of maximum activity varied from 7 to 23 years, the average being 13⅓ years. Thus, Omori concludes that in Japan the maximum epochs of destructive activity recur on an average every 13 or 14 years.

Even in the great earthquakes of the whole world there are also, as Milne points out, alternations of activity and repose. Taking into account only those earthquakes of the years 1899–1908 which were registered over the whole world or over a hemisphere, he shows that the number of earthquakes in a group varies from 2 or 3 to 15, in two cases, however, rising to 46 and 51. Such groups usually last from one to three days and are seldom prolonged over six days.

Assigning to each earthquake an intensity 2 or 1 according as it was registered over the whole world or over a hemisphere, Milne obtains the intensity of a group by adding together the intensities of all earthquakes contained in it. Then, dividing the groups into classes according to the number of days of rest

* *Journ. Coll. Sci.*, Imp. Univ. Tokyo, vol. 11, 1899, pp. 315–388, 402–403; J. Milne, *Rep. Brit. Ass.*, 1911, pp. 649–740.

which follow them, he finds the average intensity of the groups for each of such intervals. The interval between the centres of successive groups usually varies from 15 to 50 days, and is roughly proportional to the intensity of the earlier group, a group of great intensity being followed by a long period, and one of low intensity by a short period, of quiescence*.

189. Double and Multiple Earthquakes. In addition to this tendency to clustering, great earthquakes, as Milne has shown, often occur in pairs and even in triplets in widely separated districts. A typical example is that of Apr. 19, 1902. At 2.21 p.m., a great earthquake destroyed many towns and villages in Guatemala. The seismographs at Capetown, Calcutta, Bombay, and other places record its vibrations, but they also reveal the waves of a second earthquake which, according to Milne, originated at about 2.34 p.m. in the Indian Ocean in 35° S. lat., 60° E. long., approximately. The difference in time between the two earthquakes is thus about 14 minutes and the distance between their epicentres is 146°, the time required by the first preliminary tremors to traverse this arc being about 17½ minutes.

Again, on Aug. 17, 1906, a great earthquake occurred at 0 h. 8 m. or 0 h. 11 m. p.m., its epicentre lying in 31° N. lat., 168° E. long. On the same day, at 0 h. 41 m. p.m., Valparaiso and the neighbouring towns were ruined. The interval between the two earthquakes was thus about 33 minutes. To traverse the distance of 122° between the two epicentres, the first series of preliminary tremors would require nearly 16 minutes and the second series about 29 minutes.

During the years 1899–1906, there were, as Milne points out, 15 cases of great double earthquakes and 2 of triple earthquakes, the intervals between them varying from 2 to 106 minutes, with an average of 27 minutes. As the average interval between successive great earthquakes amounts to nearly a week, it is thus difficult to suppose that, in these pairs, the occurrence of the second earthquake was independent of that of the first†.

* F. Omori, *Journ. Coll. Sci.*, Imp. Univ. Tokyo, vol. 11, 1899, pp. 410–412; J. Milne, *Rep. Brit. Ass.*, 1910, pp. 54–55; 1912, pp. 92–94.

† J. Milne, *Rep. Brit. Ass.*, 1911, pp. 32–35. Milne suggests that every great earthquake causes a relief of seismic strain throughout the world, and that the disturbance in the second focus results from the arrival of the waves from the first. In about one-half the pairs, however, the

190. Synchronous Variations of Frequency in Different Districts. Milne has shown that, in some widely separated regions, the variations of frequency may be synchronous or nearly so. Taking the East Pacific districts *A*, *B* and *D* (sect. 177) and the West Indies *C* in one group, and the West Pacific districts *E* and *F*, and the Himalayas *K* in the other, the following table gives the totals for the years 1899 to 1907:

	1899	1900	1901	1902	1903	1904	1905	1906	1907	Total
E. Pacific	31	20	18	18	13	0	6	13	14	133
W. Pacific	30	16	27	45	41	28	40	53	38	318

Thus, the greater seismic activity has been manifested on the Asiatic side of the Pacific Ocean and especially in the Malay Archipelago, while the west side of South America has been least frequently disturbed. Also, since 1902, the totals for the two great regions rise and fall together in successive years.

The synchronism, however, appears to extend to still more widely separated regions, such as the Italian peninsula and Sicily; Japan, Formosa and the Philippines; North, Central and South America; and China. Considering destructive earthquakes only for the 200 years 1700–1899, Milne divides the whole period into six intervals, the first from 1700 to 1734, and each of the others of 33 years each. Any given year is considered to be one of activity or quiescence if the number of destructive earthquakes in any one region be above or below the average for the 33 years in which it occurs. Now, all four districts have shown abnormal seismic activity in 12 years and comparative quiescence in 15 years; three districts have been unusually active and one quiescent in 46 years, and three quiescent and one active in 58 years. Thus, in 131 years out of 200, or roughly in two out of three, seismic activity or quiescence has generally prevailed in these four widely separated regions of the world. It would seem from this that the seismic activity of a district does not depend only on local conditions, but may be partly

interval between the earthquakes is actually less than the time required by the first preliminary tremors to traverse the intervening distance. Is it possible that these double earthquakes are analogous to the twin earthquakes described in sects. 242–244, and that they originate in a similar manner?

governed by internal or external agencies which are general to large and widely separated areas, if not to the whole world*.

191. Secular Variation of Frequency. Every prolonged catalogue shows, not only fluctuations in frequency, but also an apparent general increase of frequency with the lapse of time. Even in a catalogue of destructive earthquakes only, this increase is manifest. In Milne's great catalogue, the average number entered during each of the first ten centuries is 17, the total number for the nineteenth century is 423, or 25 times as great†. There can be no doubt that the increase in the records is very largely due to more complete organisation and to more careful study. At the same time, it is possible that the increased frequency may be partly real. Earthquakes which devastate whole districts are events of historical importance, and thus the list for the three centuries from 1300 to 1600 should be approximately as complete as that for subsequent centuries. Now, the number of earthquakes of this class recorded in each of the seven half-centuries from 1300 to 1650 was either two or three. In the interval 1650–1700 it rises to 10, and in successive half-centuries ending with 1850 to 15, 17 and 30. It is therefore possible, as Milne suggests, that there may have been a marked increase in seismic activity from about 1650; and it is worthy of notice that this increase is shared by destructive earthquakes of less intensity and also by volcanic eruptions‡.

Periodicity of Earthquakes

192. The fluctuations in frequency noticed above are apparently not periodic. There are, however, other fluctuations, the maxima of which recur at regular intervals. The existence of some of the periods investigated—as, for example, those which may be associated with the attraction of the sun and moon—must be regarded as still unproven, the effects of such attraction being too slight to render their detection certain. On the other

* *Rep. Brit. Ass.*, 1909, pp. 56–58; 1911, pp. 36–38.

† Partly, of course, this is due to the discovery of new lands or the increasing civilisation of old ones. Thus the American records begin with the year 1520 (except for a few compiled from an old chronicle). The Philippine Islands do not contribute to the list until the year 1600, India until 1668, and New Zealand until 1848.

‡ *Rep. Brit. Ass.*, 1908, pp. 78–80; 1911, pp. 649–740.

hand, the annual variation in seismic frequency is very pronounced in nearly every country in the world. There is also evidence in some districts of a diurnal period. These two periods will now be considered. Reference will also be made to a possible fluctuation in frequency connected with the displacement of the earth's axis*.

Two remarks should perhaps be made at this stage with regard to the periodicity of earthquakes:

(i) Taking, for example, the annual period with its maximum epoch in winter, it is not suggested that earthquakes are more frequent every winter than in the preceding or following summer, but only that, in a long series of years, more earthquakes are felt during the winter than during the summer months.

(ii) Whatever the cause of a seismic period may be, that cause does not originate the earthquakes, but merely precipitates the time of their occurrence. Under the continually growing stresses to which it is subjected, the earth's crust in a given region may be on the point of displacement, and an increase of, say, the barometric pressure may be the last straw which decides *when* the displacement shall occur.

193. The simplest method of considering the periodicity of earthquakes is to represent their variation in frequency by means of curves. For the annual period, the earthquakes are usually grouped in monthly intervals, and, for the diurnal period, in hourly intervals. In Fig. 76, for instance, are given the curves representing the monthly numbers of shocks in two adjoining districts of Europe during the years 1865–1884, the continuous line corresponding to the earthquakes of Austria and the broken line to those of Hungary, Croatia and Transylvania. While varying in details, both curves indicate a pronounced maximum in winter and a minimum during the summer months. In Fig. 77

* On the periodicity of earthquakes, the following papers may be consulted:

1. Davison, C. (1). On the annual and semi-annual seismic periods. *Phil. Trans.*, vol. 184 A, 1893, pp. 1107–1169.

2. —— (2). On the diurnal periodicity of earthquakes. *Phil. Mag.*, vol. 42, 1896, pp. 463–476.

3. Knott, C. G. Earthquake frequency. *Trans. Seis. Soc. Japan*, vol. 9, 1886, pp. 1–20.

4. Schuster, A. On lunar and solar periodicities of earthquakes. *Proc. Roy. Soc.*, vol. 61, 1897, pp. 455–465.

are given the curves representing the hourly numbers of shocks in two widely separated districts, the continuous line corresponding to the earthquakes of Switzerland in the years 1878–1886, and the broken line to those of the West Indies during the historic period. Both curves, and indeed all curves deduced from non-instrumental catalogues, exhibit two apparent maxima of diurnal frequency, one shortly before midnight and the other a few hours after. There can be little doubt that the increased fre-

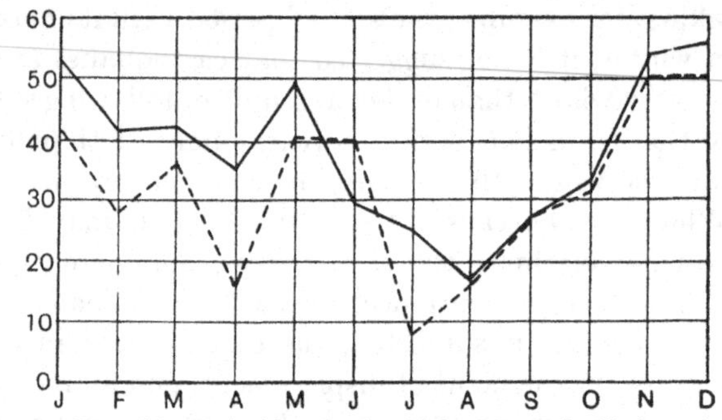

Fig. 76. Monthly variation of seismic frequency in (i) Austria and (ii) Hungary, Croatia and Transylvania.

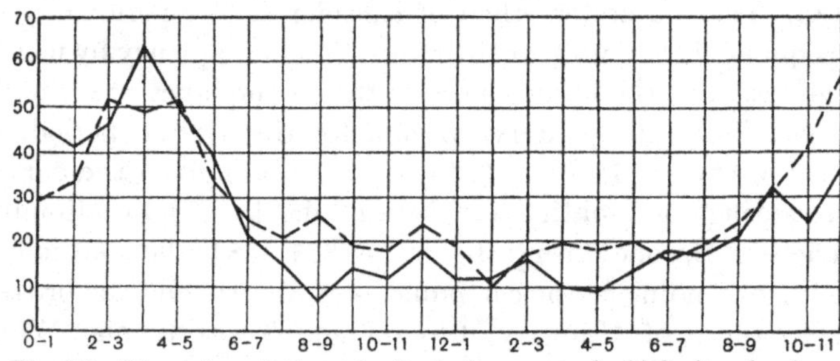

Fig. 77. Diurnal variation of seismic frequency in (i) Switzerland and (ii) the West Indies.

quency at these times is due to the favourable conditions for observation which prevail during the early hours of the night, and again about 2 or 3 a.m. when many persons lie awake in a nervous condition after their first sleep.

Curves, such as those of Figs. 76 and 77, represent the annual and diurnal variations in frequency, so far as the catalogues on which they are based are trustworthy. It does not follow, however, that they give correctly the maximum epochs of the annual and diurnal periods, for the numbers which the curves represent may be due to co-existent periods of various lengths. There may, for instance, be a semi-diurnal, as well as a diurnal, seismic period, and the curves in Fig. 77 may be the resultant of these two, and possibly of other, periods.

194. Let us suppose that the numbers of earthquakes during successive hours of the day are the sums of corresponding numbers in the following series:

4, 13, 29, 50, 73, 100, 127, 150, 171, 187, 196, 200,
196, 187, 171, 150, 127, 100, 73, 50, 29, 13, 4, 0,

and

6, 25, 50, 75, 93, 100, 93, 75, 50, 25, 6, 0,
6, 25, 50, 75, 93, 100, 93, 75, 50, 25, 6, 0.

If the numbers in the first series be plotted, the curve so obtained (represented by the broken line in Fig. 78) is the curve

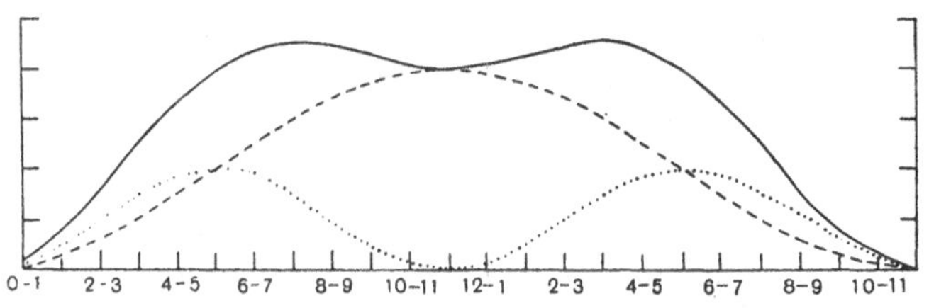

Fig. 78. Harmonic curves and their compound.

of sines, the period being one day, and the maximum epoch being at the middle of the hour 11–12 a.m., or 11.30 a.m. If the numbers in the second series be plotted, the curve so obtained (represented by the dotted line in Fig. 78) is again the curve of sines, but its period is 12 hours, and the maximum epochs occur at 5.30 a.m. and p.m. Adding together the corresponding numbers in the above two series, we obtain the following series:

10, 38, 79, 125, 166, 200, 220, 225, 221, 212, 202, 200,
202, 212, 221, 225, 220, 200, 166, 125, 79, 38, 10, 0.

The curve representing these numbers is the continuous curve in Fig. 78, in which there are two maxima, at 7.30 a.m. and 3.30 p.m., with a slightly marked minimum at 11.30 a.m. and a pronounced minimum at 11.30 p.m.

195. Method of Analysis. The variation in earthquake-frequency may be irregular, as shown by the curves in Fig. 77, or it may rise and decline gradually, as indicated by the continuous curve in Fig. 78. But, whatever the curve of variation may be, it is always possible to represent it by adding together the ordinates of a series of sine-curves with periods of respectively 1, $\frac{1}{2}$, $\frac{1}{3}$, $\frac{1}{4}$, etc., that of the period investigated, and the object of harmonic analysis is to sift out each of these periods from the others and to ascertain its amplitude and maximum epoch.

In determining these elements of the various seismic periods, the ordinary method of harmonic analysis may be used, and the results may be calculated to any degree of accuracy desired. It is important, however, to avoid giving an appearance of precision where no real precision is attainable. Only those catalogues which are based on instrumental records can lay any claim to completeness, and but few of these cover a period of years sufficient to neutralise the tendency of earthquakes to occur in groups. Thus, until our materials approach completeness, the somewhat rough form of harmonic analysis known as the *method of overlapping means* gives results which are quite as accurate as our materials warrant. This method will now be described*.

196. For the annual period, the numbers of shocks in each month are counted, and these numbers are taken to represent the rate of occurrence at the middle of the month. Six-monthly means are then calculated for each month, the mean for January being that of the six months from November to April inclusive. The middle of that period being at the end of January, it follows that the mean so obtained corresponds to the end of the month. The effect of taking six-monthly means is to eliminate the semi-annual period, if there be one, and to diminish or eliminate all other minor periods, and thus to extricate the annual period almost entirely free from others which are not being examined. To compare the results for any district with those of other districts, each six-monthly mean should be divided by the average of all twelve means, that is, by the average monthly number of earthquakes. The method of overlapping means, however, reduces each such mean in the ratio 1·589 to 1, and the difference between each reduced six-monthly mean and unity must therefore be augmented in this ratio.

* Knott, pp. 8–10, 20; Davison (1), pp. 1108–1115.

For the semi-annual period, the method is the same, except that the numbers of shocks in successive half-months are counted, and those for the first halves of January and July are added together, and so on.

The same methods are followed for the diurnal and semi-diurnal periods, except that, for the former, it is necessary to take twelve-hourly means. To obtain that for the hour 0–1 a.m., the mean of the 12 hours from 7–8 p.m. to 5–6 a.m. is calculated, and is taken to correspond to the middle of the interval, namely, to 1 a.m. The augmenting factor in this case is 1·522.

197. Let us apply this method to the third series of numbers in sect. 194. Each number in this series is the sum of the corresponding numbers in the first and second series. The former have a maximum at 11.30 a.m., the latter have maxima at 5.30 a.m. and p.m. The amplitudes of the two series of numbers are respectively 100 and 50, the ratios of which to the average (150) of the third series of numbers are $\frac{100}{150}$ and $\frac{50}{150}$ or ·67 and ·33.

Adding together the numbers for the 12 hours from 7 p.m. to 7 a.m., 8 p.m. to 8 a.m., and so on, we obtain the following twelve-hourly means (each multiplied by 12) corresponding to the middle of each 12 hours, that is, to 1 a.m., 2 a.m., etc.:

1090, 1190, 1332, 1506, 1698, 1898, 2090, 2264, 2406, 2506, 2560, 2560, 2506, 2406, 2264, 2090, 1898, 1698, 1506, 1332, 1190, 1090, 1036, 1036.

Dividing each of these numbers by their average, 1798, and multiplying the difference between each and unity by the augmenting factor for twelve-hour intervals, 1·522, we obtain the following numbers:

·37, ·46, ·59, ·73, ·91, 1·09, 1·27, 1·41, 1·54, 1·63, 1·66, 1·66, 1·63, 1·54, 1·41, 1·27, 1·09, ·91, ·73, ·59, ·46, ·37, ·33, ·33.

These numbers represent the ratio of the numbers of earthquakes during each hour to the average hourly number, so far as they are due to the diurnal period. As the highest number, 1·66, corresponds to 11 a.m. and noon, we conclude that the maximum of the diurnal period occurs at 11.30 a.m. and that its amplitude is ·66, or slightly more.

For the semi-diurnal period, the numbers for corresponding hours, a.m. and p.m., are added together, namely,

212, 250, 300, 350, 386, 400, 386, 350, 300, 250, 212, 200.

Adding together the numbers for the 6 hours from 10 to 4, 11 to 5 and so on, we obtain the following six-hourly means (each multiplied by 6), corresponding to the middle of each 6 hours, that is, to 1, 2, etc., a.m. and p.m.:

1524, 1698, 1898, 2072, 2172, 2172, 2072. 1898, 1698, 1524, 1424, 1424.

Dividing each number by the average of all, 1798, and multiplying the differences between each and unity by the augmenting factor for six-hour intervals, 1·589, we obtain the following numbers:

·76, ·92, 1·08, 1·24, 1·33, 1·33, 1·24, 1·08, ·92, ·76, ·67, ·67.

These numbers represent the ratio of the number of earthquakes during each hour, so far as they are due to the semi-diurnal period. As the highest number, 1·33, corresponds to 5 and 6 a.m., we conclude that the maxima of the semi-diurnal period occur at 5.30 a.m. and p.m. and that the amplitude of the period is ·33 or slightly more.

For the eight-hour period, the numbers for 0–1 a.m., 8–9 a.m. and 4–5 p.m. are added together, and so on. We thus obtain the following numbers:

451, 450, 447, 450, 447, 450, 451, 450.

For the six-hour period, the numbers for 0–1 a.m., 6–7 a.m., 0–1 p.m. and 6–7 p.m. are added together, and so on, and we obtain the numbers:

598, 600, 600, 600, 598, 600.

The numbers in each series are practically equal, and this shows that both the eight-hour period and the six-hour period are absent from the variation in frequency considered. The analysis has thus sifted out with accuracy the required component periods.

198. It is important that the catalogues which are subject to analysis should be as complete as possible and should cover a long series of years. If the total number of earthquakes included should be small, the analysis of the figures may provide a period which does not really exist. Schuster has shown that, if the earthquakes should occur at random, harmonic analysis may give rise to an apparent seismic period of amplitude $\sqrt{(\pi/n)}$, where n is the number of earthquakes. Moreover, earthquakes occur, not at random, but in groups. Thus, in any isolated record, unless the amplitude be much in excess of the above amount or expectancy, the period indicated by the analysis cannot be regarded as established. If, however, the maximum epochs of a particular period should agree approximately in neighbouring districts or for the same district in different intervals of time, the existence of the period may be regarded as probable even if the calculated amplitude were to fall below the expectancy.

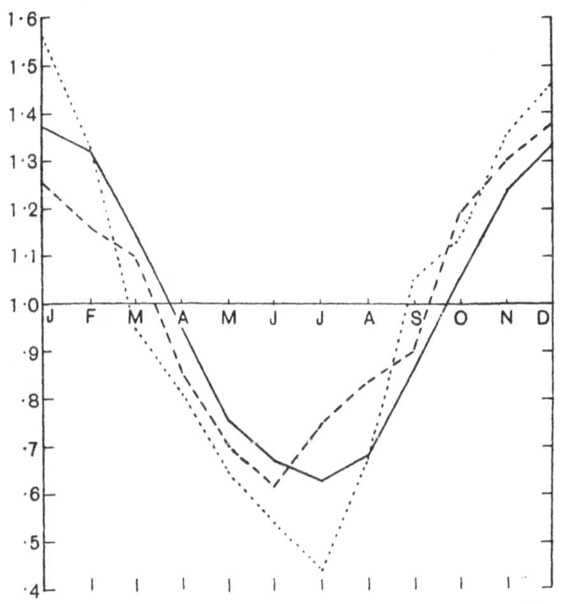

Fig. 79. Annual periodicity of earthquakes in (i) Austria, (ii) Hungary, Croatia and Transylvania, and (iii) Switzerland and the Tyrol.

For example, in Fig. 79, there are shown the curves representing the annual period in three neighbouring countries for the years 1865–1884. The continuous curve is that for Austria, the amplitude being ·37, the number of earthquakes 461, and the expectancy ·08. The broken line represents the annual period for Hungary, Croatia and Transylvania, the amplitude being ·31, the number of earthquakes 384, and the expectancy ·09. The dotted line shows the same period for Switzerland and the Tyrol, the amplitude being ·56, the number of earthquakes 524 and the expectancy ·08. Even if the amplitude had in each case fallen below the expectancy, the fact that the maximum epochs occur at the end of January, December and January would support the reality of the period.

199. Annual Seismic Period. The method of analysis described above was applied by Knott to six catalogues in 1884, and by Davison some years later to a much larger number. Many of the latter, however, give rise to an amplitude which is either less than the expectancy or not much in excess. Excluding all those cases in which the calculated amplitude is less than three times the expectancy, we have the following results, the maximum epoch in each case being at the end of the month mentioned:

District	Catalogue	No. of earthquakes	Expectancy	Ann. Seis. Period	
				Max. Epoch	Ampl.
Northern Hemisphere ...	Mallet	5879	·02	Dec.	·11
" "	Fuchs	8133	·02	Dec.	·29
Europe	"	5499	·02	Dec.	·35
"	Perrey	1961	·04	Dec.	·22
Great Britain	D. Milne	205	·12	Nov.	·49
France	Fuchs	297	·10	Dec.	·41
"	Perrey	656	·07	Jan.	·33
Austria	Fuchs	461	·08	Jan.	·37
Hungary, Croatia and Transylvania	"	384	·09	Dec.	·31
Switzerland and the Tyrol ...	"	524	·08	Jan.	·56
Basin of the Rhone ...	Perrey	184	·13	Nov.	·46
" " Rhine	"	529	·08	Jan.	·38
South-east Europe	Schmidt	3470	·03	Dec.	·21
Balkan Peninsula	Fuchs	1326	·07	Dec.	·27
Algeria	"	135	·15	Dec.	·67
Asia	"	458	·08	Feb.	·33
Caucasia	"	152	·14	Jan.	·56
Tokyo	Milne	1104	·05	Feb.	·19
North America	Fuchs	552	·08	Nov.	·35
New England	Brigham	212	·12	Dec.	·51
California	Holden	949	·06	Oct.	·30
West Indies	Fuchs	205	·12	Oct.	·59
Hawaii	"	245	·11	June	·33
New Granada and Venezuela	"	272	·11	Feb.	·64
Southern Hemisphere ...	"	751	·07	Aug.	·37
N.S. Wales, Victoria and S. Australia	Hogben	159	·14	May	·48
Chili	Knott	212	·12	Aug.	·48
Peru, Bolivia and Quito ...	Fuchs	350	·09	July	·48

Thus, of the 25 districts which lie in the northern hemisphere, the maximum epoch occurs at the end of November in 3 cases, December in 10, and January in 5 cases. Of the four which lie in the southern hemisphere, the maximum epoch occurs at the end of May in New South Wales, etc., July in Peru, etc., and August in the southern hemisphere and in Chili. In other words, the maximum epoch occurs as a rule in winter in both hemispheres.

In the semi-annual period, the amplitude is less than three times the expectancy in all but 13 records. The average value in these cases is ·29 and the maximum value ·59. The maximum epochs occur in different months, though, in a limited region, such as the Japanese Empire, they are confined almost entirely to those of spring and autumn.

200. Origin of the Annual Seismic Period. In nearly every part of the world, there is an annual barometric period, the amplitude of which varies from $\frac{1}{20}$ to $\frac{1}{2}$ an inch. In Europe, the maximum is usually in November, over Asia in December, throughout North America in November, December or January, and in Hawaii in April and May. In the southern hemisphere, the maximum occurs as a rule in the corresponding months—in New Zealand in April, south-east Australia in May or June, and South America in June or July. Thus, as a general rule, the epoch of the seismic maximum either occurs in the same month as the barometric maximum or follows it by a month or two.

G. H. Darwin has estimated the distortion of the earth's surface under parallel waves of barometric elevation and depression. Taking 5000 miles as the wave-length of the barometric undulation, and 5 cms. as the difference in pressure below the crest and trough of the undulation, Darwin calculates that the ground will be 9 cms. higher under the barometric depression than under the elevation. Moreover, the vertical distortion decreases very slowly as the depth increases, being practically the same at the surface and at a depth of 50 miles.

Now, since the movement which produces even a moderately strong earthquake is slight in vertical extent; since, moreover, the work to be done is, not the compression of the solid crust, but the slight depression of a fractured mass of rock, the support of which is nearly but not quite withdrawn, it seems possible that the annual variation in barometric pressure, small as it is, may be competent to produce the annual variation in seismic frequency.

A curious test of this explanation is given by the small amplitude of the seismic period in some insular districts—for instance, ·10 in Zante, ·08 in Japan, ·05 and ·06 in New Zealand. In these districts, many of the earthquakes are of submarine origin, and, as the barometric pressure changes throughout the year, the sea has time to take up its equilibrium position, so that the total change of pressure on the sea-bed is much less than that on land*.

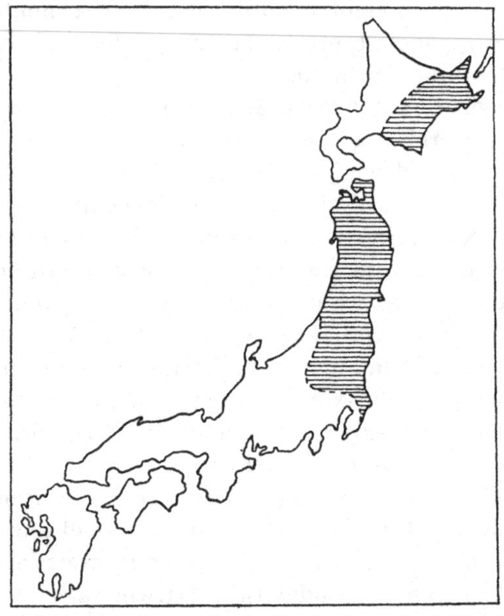

Fig. 80. Variation of annual seismic maximum-epochs in Japan.

201. To ascertain how far the small amplitude of the annual period in Japan may have such an origin, Omori has examined the seasonal distribution of earthquakes at 26 stations in the empire. At 15 stations he finds that the annual maximum occurs in winter or spring, and at 11 in summer or autumn. At nine stations, however, the records include clusters of after-shocks, and at three others the maximum epoch is doubtful

* In a large insular region like Japan, the maximum epoch occurs during the winter months in some districts and during the summer months in others. The effect of this would naturally be to diminish the amplitude of the annual period for the whole region.

owing to the brevity of the record. The application of the method of overlapping means to the remainder shows that the maximum epoch occurs in winter at seven stations, and in summer at seven others. The stations with a summer maximum lie within the shaded area in the sketch-map (Fig. 80)*.

Now, most of the earthquakes which disturb the districts with a winter maximum have an inland origin, and their frequency is probably governed by the variation in barometric pressure, which has its maximum in winter. On the other hand, the majority of the earthquakes which disturb the north-east of Japan (represented by the shaded area) originate under the Pacific. Now, at two stations along this portion of the coast, the sea-level is lowest in February, and higher in September by 276 mm. at one station and 217 mm. at the other. The barometric maximum occurs in November and January, the total range throughout the year being 9·3 mm., which corresponds to 126 mm. of water. Thus, the sea-bed is subjected to a greater total pressure in the summer months than in February–April, and this, according to Omori, is probably the cause of the summer maximum of the annual seismic period in the north-east of Japan†.

202. Periodicity in relation to Intensity. If periodicity be due to causes which precipitate, rather than produce, earthquakes, we should expect the periodicity of very slight shocks to be more marked than that of very strong earthquakes.

In his catalogue of recorded earthquakes, Mallet gives the times of occurrence of 187 very slight shocks and 641 destructive earthquakes (the expectancy being ·13 and ·07 respectively). The maximum epoch of the slight shocks occurs at the end of June (amplitude ·16), and of the destructive earthquakes at the end of December (amplitude ·13).

This reversal of epoch is also shown in other records. The Japanese earthquakes of 1885–1892 may be divided into slight, moderate, and strong shocks, according as the disturbed area

* The boundary of the shaded area in Fig. 80 differs slightly, owing to the mode of treating the statistics, from that laid down by Omori. Three small detached areas with a summer maximum are also omitted, owing to the exclusion of the records from the twelve stations mentioned above.

† F. Omori, *Publ. Eq. Inv. Com.*, No. 8, 1902, pp. 1–94; No. 18, 1904, pp. 23–26; *Bull. Eq. Inv. Com.*, vol. 2, 1908, pp. 35–50; vol. 5, 1913, pp. 39–86.

is less than 600 sq. miles, between 600 and 6000, or more than 6000 sq. miles, with the following results:

	No. of earth-quakes	Expect-ancy	Annual Period	
			Max. epoch	Ampl.
Slight	2256	·04	Middle of Oct.	·14
Moderate	567	·07	Middle of Mar.	·17
Strong	176	·13	End of Mar.	·17

Three different catalogues of the earthquakes of Zante give similar results, the third catalogue differing from the second merely in including a larger number of very slight shocks*.

Catalogue	No. of earth-quakes	Expect-ancy	Annual Period	
			Max. epoch	Ampl.
Schmidt and Fuchs	246	·11	End of Dec.	·29
Barbiani (1)	1326	·05	End of Aug.	·10
,, (2)	1663	·04	,, ,,	·29

203. Diurnal Seismic Period. In the analysis of the diurnal periodicity of earthquakes, it is essential that the records used should be entirely instrumental (sect. 193). Such records are few in number and not of long duration. The method of harmonic analysis has been applied to those of Tokyo, Japan, Manila, and various Italian observatories. The following table gives the results obtained by the method of overlapping means:

District	No. of earth-quakes	Expec-tancy	Diurnal		Semi-diurnal	
			Max. epoch	Ampl.	Max. epoch a.m. & p.m.	Ampl.
Tokyo, whole year	2539	·04	11 a.m.	·08	9	·10
,, winter	1290	·05	12 noon	·06	9	·13
,, summer	1249	·05	11 a.m.	·10	9	·06
Oita	237	·12	11 a.m.	·39	8	·12
Japan	1175	·05	12 noon	·11	irregular	·05
Manila	208	·12	11 a.m.	·30	3	·19
Italy	8177	·02	1 p.m.	·32	11	·11

* Davison (1), pp. 1116–1120. According to Omori (*Bull. Eq. Inv. Com.*, vol. 2, 1908, pp. 17–20), the same reversal characterises the weak and strong earthquakes of Tokyo and Kyoto.

The diurnal period is thus less clearly marked than the annual period. In most cases, the amplitude does not greatly exceed the expectancy. The effect of the tendency to occur in groups is; however, slighter than in the annual period; and there is also a close agreeent in the maximum epoch, not only in different districts, but in the same district during the winter and summer months. So far as we may judge from the records at our disposal, there seems to be a diurnal period with its maximum between 11 a.m. and 1 p.m.

The semi-diurnal period is still less pronounced, the average amplitude being ·10 instead of ·19. The maximum epoch is variable, except in Japan, where it occurs about 9 a.m. and p.m.

204. After a great earthquake, the crust is in a peculiarly sensitive condition, and fluctuations in frequency should then be strongly marked. Instrumental records have been obtained of the after-shocks of the Mino-Owari earthquake of Oct. 28, 1891, at Gifu and Nagoya, and of the Hokkaido earthquake of Mar. 22, 1894. The results are given in the following table:

District	No. of earth-quakes	Expec-tancy	Diurnal		Semi-diurnal	
			Max. epoch	Ampl.	Max epoch a.m.&p.m.	Ampl.
Gifu:						
Oct. 29–Nov. 10, 1891	1257	·05	12 noon	·20	6	·11
Nov. 11, 1891–Dec. 31, 1899	2856	·03	0½ a.m.	·13	8½	·06
Nagoya:						
Oct. 29–Nov. 10, 1891	572	·07	3 a.m.	·35	2	·19
Nov. 11, 1891–Dec. 31, 1899	1282	·05	about midn.	·14	3	·11
Nemuro:						
Mar. 23–31, 1894	345	·10	1½ a.m.	·36	7	·25
Apr. 1, 1894–Dec. 31, 1899	646	·07	0½ p.m.	·16	5	·13

The diurnal variation in the two Nagoya records is represented in Fig. 81, in which the continuous line indicates the variation during the earlier period and the broken line that during the later.

Thus, in every case, the amplitude of the diurnal period is about three times, or more than three times, the expectancy. We may therefore conclude that: (i) after-shocks of great earthquakes are governed by a diurnal fluctuation in frequency, the maximum occurring at, or shortly after, midnight, and (ii) the diurnal period is more marked during the week or ten days following the principal earthquake than afterwards. The semi-diurnal period is less pronounced than the diurnal period, and is more marked during the week following the principal earthquake than afterwards; the maximum epochs are variable.

As the diurnal barometric period has a maximum epoch a few hours after midnight, it is possible that the diurnal variation in the frequency of after-shocks may be connected with the diurnal variation in barometric pressure*.

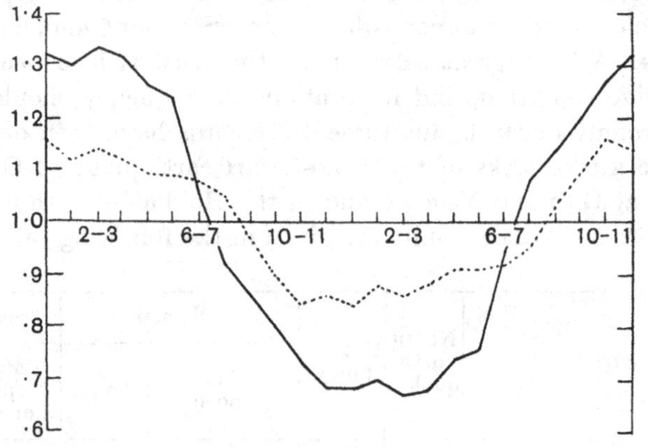

Fig. 81. Diurnal periodicity of the after-shocks of the Mino-Owari earthquake of Oct. 28, 1891, at Nagoya, (i) Oct. 29–Nov. 10, 1891, (ii) Nov. 11, 1891–Dec. 31, 1899.

205. Relation between Variations of Latitude and the Frequency of Earthquakes. The recent discovery of the variations of the Pole from its mean position has led Milne to make an interesting comparison between the frequency of great earth-

* Davison (2), pp. 465–476. Omori has examined a large number of seismic records from different observatories in Japan. Most of them are non-instrumental for part of the time. The method of analysis used by him, moreover, is practically that of taking two-hourly overlapping means—a form of smoothing which is not well adapted for the purpose in view. By combining in one diagram, the diurnal variations in earthquake-frequency and barometric pressure throughout the day, Omori is led to consider the former variation as a result of the latter (*Publ. Eq. Inv. Com.*, No. 8, 1902, pp. 53–94).

quakes and the changes in direction (or deflections) of the movements of the Pole. The comparison is made for the years 1892 to 1904. Dividing the year into ten intervals of 36½ days each, Milne finds the numbers of great earthquakes in the intervals in which deflection occurs and in those immediately before and after them. During the years considered, there were 23 deflection-periods, and in 18 of these the number of earthquakes in the deflection-period was greater than in either adjoining period. The total number of earthquakes in the deflection-periods was 287, during those preceding them 167, and during those following them 217—numbers which are as 100 : 58 : 76. Milne thus concludes that great earthquakes are frequent about the times when changes occur in the direction of the polar displacements and especially when those rates of change are rapid.

That the deflections of the polar movements govern the frequency of the stronger earthquakes only seems clear from a comparison which Omori has made for the interval from Aug. 1885 to Dec. 1903. He finds that all the destructive earthquakes (16 in number) which visited Japan during this interval occurred exactly or very nearly at those epochs when the latitude of Tokyo was at a maximum or minimum. The strong but non-destructive earthquakes (41 in number) show a similar tendency, though in a less marked degree. Of the slight or sensible earthquakes (303 in number), 180 occurred during 42 months in which the latitude was close to its maximum or minimum value, and 123 during the 28 months in which the latitude was increasing or decreasing. The corresponding rates per month were practically equal, namely, 4·3 and 4·4 respectively. Of moderately strong earthquakes, 78 occurred during 24 months of maximum or minimum latitude, and 54 during 17 months of decreasing or increasing latitude. The corresponding rates per month were again nearly equal, namely, 3·3 and 3·2. Thus, Omori concludes that destructive earthquakes in Japan have a marked tendency to occur in the epochs of maximum or minimum latitude at Tokyo, but that earthquakes of a moderate or slight degree of intensity are unaffected by the changes of latitude.

The exact relation between the changes of latitude and the earthquake-frequency (supposing further investigations should prove its reality) is at present unknown. The crust-displacements in great earthquakes are insufficient to produce the observed deflections of the Pole, and it is inconceivable that deflections so minute should be the cause of the earthquakes. On the whole, it seems probable that both are rather due to the action of some common cause*.

* J. Milne, *Rep. Brit. Ass.*, 1900, pp. 107–108; 1903, pp. 78–80; 1906, pp. 97–99; A. Cancani, *Boll. Soc. Sis. Ital.*, vol. 8, 1903, pp. 286–290; F. Omori, *Publ. Eq. Inv. Com.*, No. 18, 1904, pp. 13–21; C. G. Knott, *Rep. Brit. Ass.*, 1907, pp. 91–92.

CHAPTER XII

ACCESSORY SHOCKS

206. No great earthquakes, and few earthquakes of even moderate strength, occur alone. They are usually, but not always, preceded by slight *fore-shocks*; they are invariably followed by a number, generally a large number, of *after-shocks*. Both fore-shocks and after-shocks are confined to the region containing or immediately surrounding the epicentre of the principal earthquake. Besides these, there are others, known as *sympathetic shocks*, which occur in neighbouring districts but which are precipitated by the occurrence of the principal earthquake and by the changed stresses which it suddenly introduces in the surrounding crust*.

FORE-SHOCKS

207. Number and Intensity of Fore-Shocks. Fore-shocks, when they occur, are usually few in number and of slight intensity. Occasionally, they attain a semi-destructive strength. The great earthquake of central Japan, which occurred on July 9, 1854, was preceded on the 7th inst. by two shocks at 1 p.m. and 2 p.m., strong enough to crack some plastered walls within the epicentral area of the principal earthquake. They were followed by incessant earth-sounds and by at least 27 minor shocks the same day. On July 8, the earth-sounds continued, and there were a few small shocks at about 8 p.m. Six hours later the great earthquake occurred.

As a rule, the fore-shocks are more insignificant. The Charleston earthquake of Aug. 31, 1886, was preceded on Aug. 27, at about 8 a.m., by a slight earthquake felt at Summerville, a village 22 miles to the north-west and not far from the principal epi-

* On the accessory shocks of earthquakes, the following papers may be consulted:

1. Davison, C. On the distribution in space of the accessory shocks of the great Japanese earthquake of 1891. *Quart. Journ. Geol. Soc.*, vol. 53, 1897, pp. 1–15.
2. Omori, F. On the after-shocks of earthquakes. *Journ. Coll. Sci.*, Imp. Univ. Tokyo, vol. 7, 1894, pp. 111–200.

centre. On Aug. 28, at 4.45 a.m., another and stronger earthquake occurred at the same place and was felt as far as Charleston. During this and the following day, several other slight shocks were noticed at Summerville, but were succeeded by an interval of repose which lasted until 9.51 p.m. on Aug. 31, when the great earthquake occurred. Again, the first undoubted fore-shock of the Riviera earthquake of 1887, occurred on Feb. 22 at 8.30 p.m., about 10 hours before the principal earthquake, and this was followed by five others, one of which was perceptible all over the Riviera, and in Piedmont and Corsica. The principal earthquake occurred on Feb. 23, at 6.20 a.m.

There appears to be no relation whatever between the occurrence of fore-shocks and the strength of the principal earthquake. British earthquakes, which are never of great strength, are sometimes preceded by perceptible shocks. Two fore-shocks of the Hereford earthquake of 1896, for instance, disturbed areas of 6300 and 6400 square miles, or about one-fifteenth the area shaken by the principal earthquake. On the other hand, the Messina earthquake of 1908 and the Californian earthquake of 1906, with its great fracture 270 miles in length, took place suddenly, without warning of any known kind. The Assam earthquake of June 12, 1897, one of the first rank of earthquakes with an epicentral area of 6000 or 7000 square miles, was preceded, if at all, by tremors and earth-sounds so faint that, but for the subsequent earthquake, they would have passed unrecorded.

It is possibly from the absence of seismographs or from want of careful observations, that fore-shocks seem to be rare or so often escape notice. Thus, it is said that the great Mino-Owari earthquake of 1891 was heralded by a strong shock 58 hours earlier, though earth-sounds were heard within the epicentral area during the intervening days. A more detailed examination of the previous earthquakes of the district shows, however, that a preparation for the principal earthquake had been going on for several years.

208. Distribution in Time of the Mino-Owari Fore-Shocks. The Mino-Owari earthquake occurred on Oct. 28, 1891, at 6.37 a.m. The extraordinary fault-scarp formed on that occasion is described in sect. 86. Its course is represented by the continuous line in the accompanying map (Fig. 82). The dotted lines indicate the boundary of the meizoseismal area. The area shown in the map is bounded by the parallels of 34° 40′ and 36° 20′ N. lat., and by the meridians of 2° 10′ and 3° 50′ W. long. of Tokyo, so that, if the area be divided by N.–S. and E.–W. lines one-sixth of a degree apart, there are ten rectangles adjoining each boundary of the map and 100 rectangles within its

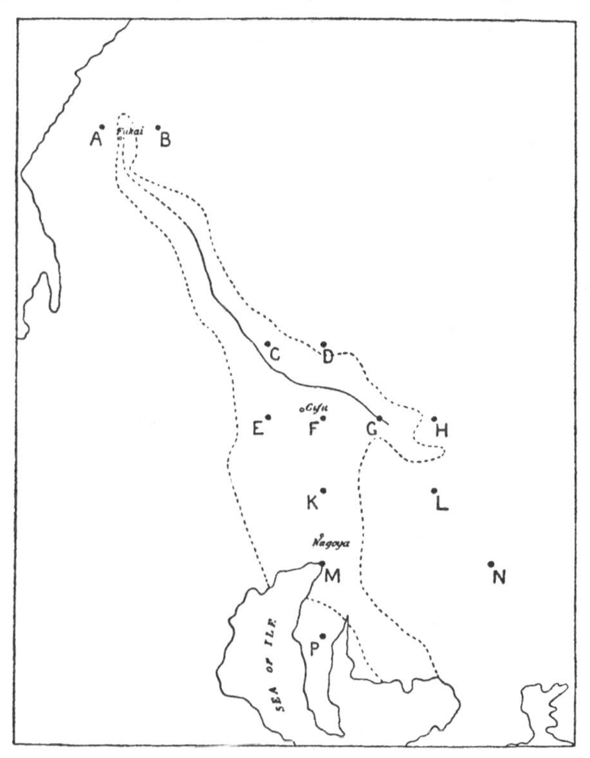

Fig. 82. Course of the fault-scarp, etc., of the Mino-Owari earthquake of 1891.

area. The points *A–P* denote the centres of the rectangles in which the majority of the fore-shock and after-shock epicentres lie. The points *C* and *D*, and possibly also *E* and *F*, seem to be connected with the main fault; *A* and *B* with the northern end of its fault-scarp, *G* and *H* with the southern end, and *L* and *N* with a probable continuation of the fault to the south-east. The central points *K*, *M* and *P* are probably connected with a deep-seated fault which follows the main branch of the meizoseismal area to the south, but of which no trace as a scarp is visible at the surface.

The number of epicentres within the whole area of the map for the different years from 1885 to 1891 (to Oct. 27) varies from 16 to 52, and the number within the rectangles *A–P* from 4 to 32. Taking account of the fact that there are 13 rectangles along the course of the fault and 87 rectangles elsewhere, it follows that in 1885 there were 5·4 times as many centres near the faults as in an equal area elsewhere, in 1886 3·9 times, and in 1887 2·2 times. It is possible that this decline in relative frequency marks the decadence from the last strong earthquake in the Mino-Owari district, which occurred in 1859. After 1887, however, there is a rapid increase in the ratio. In 1888, it rises to 5·5, in 1889 to 7·0, and during the interval from Jan. 1, 1890, to Oct. 27, 1891, it rises still further to 10·5. The first symptom of the coming of the great earthquake is therefore an increase in the frequency of earthquakes along the lines of fault relatively to the frequency elsewhere*.

209. Distribution in Space of the Mino-Owari Fore-Shocks. Still more significant is the distribution of the fore-shocks of the Mino-Owari earthquake. During the five years 1885–1889, the deep-seated fault was almost entirely inactive, and, along the main fault, the epicentres were chiefly confined to the central region. During the 22 months before the earthquake, however, a remarkable change occurred in the distribution of the epicentres. This is represented in Fig. 83. The central region of the main fault is still the principal seat of activity, but the curves now follow the courses of both faults. Earthquakes indeed occurred along the whole fault-system, especially along the line of the deep-seated fault and the continuation of the main fault towards the south-east. Moreover, they occurred with some approach to uniformity along the whole fault-region; the marked concentration of activity which characterises the after-shocks (sect. 215) is hardly perceptible among the fore-shocks†.

210. Prevision of Earthquakes by means of Fore-Shocks. With our present knowledge of the preparation for the Mino-Owari

* Davison, pp. 9–12. † Davison, pp. 11–12.

earthquake, the phenomena by which the occurrence of the earthquake might have been foreseen are (i) the increase in frequency of fore-shocks from 1887 until October 1891 in the district surrounding the fault-system as compared with equal areas elsewhere; but especially (ii) the outlining of the fault-system by the curves of equal frequency of fore-shocks during the two years immediately preceding the earthquake.

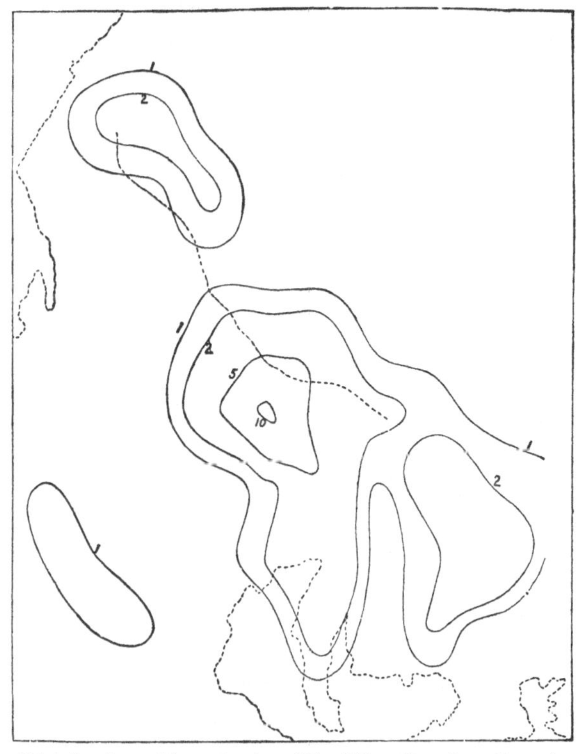

Fig. 83. Distribution of fore-shocks of the Mino-Owari earthquake of 1891.

With a sufficient number of observing stations, the location of the epicentre of an earthquake is an easy matter. It is much simpler than the measurement of the displacement of pillars built into the ground, as suggested by Reid. Moreover, it is possible that a great earthquake may, as in the Kangra earthquake of 1905, occur without very perceptible distortion of the surface-crust*.

* C. Davison, *Beitr. zur Geoph.*, vol. 12, 1912, pp. 9–15.

After-Shocks

211. Number of After-Shocks. In striking contrast with the paucity of fore-shocks is the extraordinary abundance of the after-shocks. In some cases, it is no exaggeration to say that for days the epicentral area is never actually at rest. During the night following each of the earthquakes in north-east Greece on Apr. 20 and 27, 1894, the ground within the two innermost isoseismal lines was in a state of almost incessant disturbance. After the great Assam earthquake of June 12, 1897, the ground at several places in the epicentral area remained for days in a state of tremor, interrupted frequently by sensible shocks, and occasionally by severe earthquakes. At Bordwar, the surface of a glass of water was observed for a week afterwards to be continually trembling; and at Tura a lamp was kept swinging for three or four days.

Without instrumental aid, all registers of after-shocks are inevitably incomplete. Indeed, when the trembling appears to be continual, separate after-shocks cannot be distinguished. It is only in rare cases when there are uninjured seismographs in the neighbourhood of the epicentre that we can form any idea of the actual frequency of the after-shocks. After the Mino-Owari earthquake of Oct. 28, 1891, the seismographs at Gifu and Nagoya, though overthrown, were re-erected within a few hours, the former recording 318, and the latter 185, shocks on the day after the earthquake. During the first 30 days, 1746 after-shocks were registered at Gifu, the total number by the end of 1893 amounting to 3365. After the Kumamoto earthquake of July 28, 1889, 340 shocks were recorded at Kumamoto during the first 30 days and 833 during the first two years. The Kagoshima earthquake of Sep. 7, 1893, was followed by 278 after-shocks at Chiran within the first 30 days; the Hokkaido earthquake of Mar. 22, 1894, by 431 after-shocks during the first 30 days at Nemuro (74 miles distant from the epicentre) and 715 during the first year.

Numbers approaching or exceeding those given above are the results of personal observations in several earthquakes. For instance, the Tenpo (Japan) earthquake of Aug. 19, 1830, was succeeded by 681 after-shocks at Kyoto in the first six months; the Zenkoji (Japan) earthquake of May 8, 1847, by 930 shocks at Matsushiro in the first 31 days; and the great Japanese earthquake of Nov. 4, 1854, by not less than 919 shocks before the end of the following year. In the first two of these earthquakes, the actual numbers of shocks probably rivalled that of the Gifu series of Mino-Owari after-shocks. After the Messina earthquake of Dec. 28, 1908, 949 after-shocks were counted at Messina by the end of 1909. Owing to the number and wide dispersion of independent centres of activity, the after-shocks of the great Assam earthquake of June 12, 1897, probably surpass in number those of any other known earthquake. In little more than three days, 561 shocks were felt in N. Gauhati. During one year, from Oct. 1, 1897, to Sep. 30, 1898, 1050 after-shocks were recorded at Maophlang and 841 at Mairang*.

212. Intensity of After-Shocks. The great majority of the after-shocks of an earthquake are of very slight intensity, the strongest as a rule being far inferior to the principal shock. Of the 3365 after-shocks of the Mino-Owari earthquake of 1891 recorded within a little more than two years at Gifu, 10 were violent, 97 strong, 1808 weak, 1041 feeble, and 409 were earth-sounds. The Riviera earthquake of Feb. 23, 1887, occurred at 6.20 a.m., and was followed at 6.29 a.m. and 8.51 a.m. by two shocks which added to the destruction and loss of life caused by the principal earthquake. After the great Assam earthquake of 1897, eight shocks were felt at Calcutta, 250 miles from the epicentral area, one of them beyond Allahabad to a distance of more than 550 miles. Of the Jamaica earthquake of 1907, no fewer than 148 after-shocks were strong enough to be recorded by Milne seismographs in Great Britain.

The interest of after-shocks, however, lies not in their strength but in their weakness. For the most part, they are of local origin. The shocks felt at one place are not usually the same as those felt at another only a few miles away. Thus, Maophlang and Mairang, two places in the epicentral area of the Assam earthquake (Fig. 47), are only 11 miles apart. During 17 days (Sep. 12–28, or three months after the earthquake), 92 after-

* F. Omori, pp. 111–200; *Publ. Eq. Inv. Com.*, No. 7, 1902, pp. 33–51; *Bull. Eq. Inv. Com.*, vol. 2, 1908, pp. 185–195; *Boll. Soc. Sis. Ital.*, vol. 2, 1896, pp. 152–155; Oldham, pp. 124–128, and *Mem. Geol. Surv. India*, vol. 30, 1900, pp. 1–102.

shocks were felt at Maophlang, 37 being described as smart, 45 slight and 10 feeble. In the same interval, 83 shocks were felt at Mairang, 6 being smart, 10 slight and 67 feeble. It is difficult to obtain correct time in Assam, but, regarding shocks as identical if their recorded times differ by not more than 15 minutes, there were in the interval referred to 19 shocks common to both places, leaving 73 as peculiar to Maophlang and 64 to Mairang. Moreover, of the 19 common shocks, only one was considered as smart at both places, 12 were smart at one and slight or feeble at the other, while 6 were slight at one and feeble at the other and may have been independent shocks.

Again, the after-shocks of the Mino-Owari earthquake of 1891 were recorded by seismographs at six observatories, Gifu being 17 miles from the central portion of the Neo Valley where most of the shocks originated, Nagoya 37 miles, Tsu and Kyoto 61, Osaka 88, and Tokyo 166, miles. The number of after-shocks recorded from Oct. 28, 1891 to Dec. 31, 1893, was 3365 at Gifu, 1298 at Nagoya, 314 at Tsu, 125 at Kyoto, 70 at Osaka and 30 at Tokyo. Omori has represented the local character of the after-shocks by means of curves drawn through all places at which the same number of shocks were felt. From the map for November 1891, he finds that 200 after-shocks were felt to a mean distance of 25 miles, 100 to a distance of 48 miles, and 10 to a distance of 112 miles*.

213. Decline in After-Shock Frequency. Numerous as after-shocks are for a few days after a great earthquake, their decline in frequency is at first very rapid. The daily numbers recorded at Gifu during the first seven days after the Mino-Owari earthquake of 1891 were 318, 173, 126, 99, 92, 81 and 78. A month later, the average daily number dwindled to 18. The decline in frequency of the after-shocks at Gifu is represented by the continuous line in Fig. 84, in which the numbers of after-shocks in successive months from Nov. 1891 to Dec. 1893 are represented by the distances of the small crosses from the horizontal line. Thus, in Nov. 1891, there were 1087 after-shocks, in the next

* F. Omori, *Journ. Coll. Sci.*, Imp. Univ. Tokyo, vol. 7, 1894, pp. 112–113; Oldham, p. 125; J. Milne, *Rep. Brit. Ass.*, 1908, pp. 64–66; 1909, pp. 51–55; F. Omori, *Publ. Eq. Inv. Com.*, No. 7, 1902, pp. 29–31.

month 416. In Dec. 1892, the number had decreased to 39, and in Dec. 1893 to 16.

The decline in frequency, as will be seen from the curve in Fig. 84, is far from uniform. The succession of weak and feeble shocks was broken from time to time by violent earthquakes, each of which, as on Jan. 3 and Sep. 7, 1892, was followed by its own train of after-shocks and so gave rise to temporary fluctuations in the total number. The dotted curve in Fig. 84

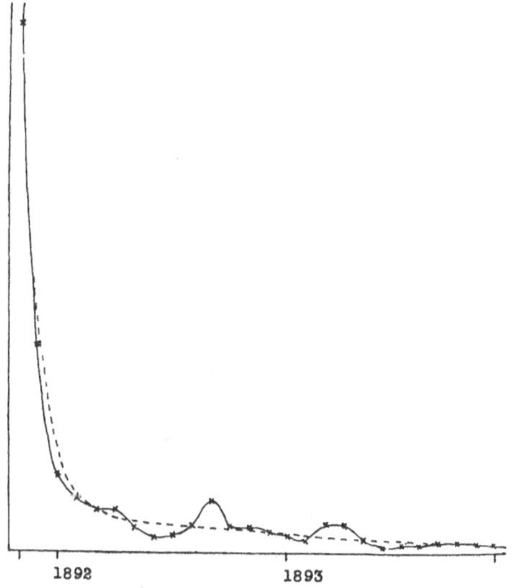

Fig. 84. Decline in frequency of after-shocks of the Mino-Owari earthquake of 1891.

is drawn so as to smooth away these irregularities. In all probability, it represents the true law of decline in frequency, so far as the original earthquake is concerned.

This dotted curve differs but little from a rectangular hyperbola. Indeed, Omori finds that, in all the earthquakes which he has examined, its equation is of the form

$$y = \frac{k}{x + h},$$

where h and k are constants, and y is the number of earthquakes felt within a given interval at time x, measured from a fixed epoch. For the Mino-Owari after-shocks recorded at Gifu, Omori uses the half-daily numbers of after-shocks during the five days Oct. 29–Nov. 2, 1891. Inserting for y the number of after-shocks during the 12 hours denoted by x, measured from the first half of Oct. 29, he thus obtains ten equations for finding h and k, the values of which are determined by the method of least squares, the resulting equation being

$$y = \frac{440{\cdot}7}{x + 2{\cdot}31}.$$

Though this equation is obtained from the numbers of after-shocks recorded during the first five days, it has nevertheless been used by Omori for determining the numbers of after-shocks felt after the lapse of six to eight years. Taking account of the usual annual number of earthquakes (18·3) recorded at Gifu, Omori estimates that the number of earthquakes in the two years 1898–1899 should be 160. The number actually recorded was 163*.

214. Decline in After-Shock Intensity. The decline in the intensity of after-shocks, like their decline in frequency, is rapid at first and fluctuating. Of the numerous after-shocks of the Assam earthquake of June 12, 1897, eight were strong enough to be felt in Calcutta, and all of them occurred within the first four months. The after-shocks of the Kumamoto earthquake of July 28, 1889, included one violent earthquake on Aug. 3, and 76 strong shocks, 29 of which occurred before the end of July, and the last on May 28, 1890. The Hokkaido earthquake of Mar. 22, 1894, was followed at Nemuro by 11 strong shocks, of which eight occurred before the end of March, two in April, and one on July 14. Of the 10 violent after-shocks of the Mino-Owari earthquake of Oct. 28, 1891, nine occurred within the first four months, and the last on Sep. 7, 1892. All of the 97 strong shocks occurred within the first 13 months, and all the weak ones but four within the first 20 months. Towards the end of 1893, besides the four weak shocks, only feeble shocks and earth-sounds were observed.

* F. Omori, p. 118, and *Publ. Eq. Inv. Com.*, No. 7, 1902, pp. 27–29. It should be mentioned that, as regards the Riviera earthquake of Feb. 23, 1887, and the Messina earthquake of Dec. 28, 1908, Cavasino and Agamennone find that the decline in frequency of the after-shocks does not follow Omori's law (*Boll. Soc. Sis. Ital.*, vol. 15, 1911, pp. 129–143; *Rivista di Astronomia*, etc., Nov. 1912). The explanation probably is that both of these earthquakes were twin earthquakes (see sect. 216).

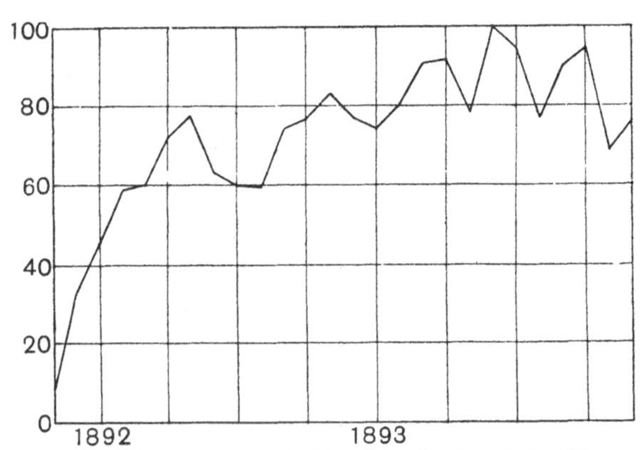

Fig. 85. Percentage of feeble after-shocks of the Mino-Owari earthquake of 1891 at Gifu.

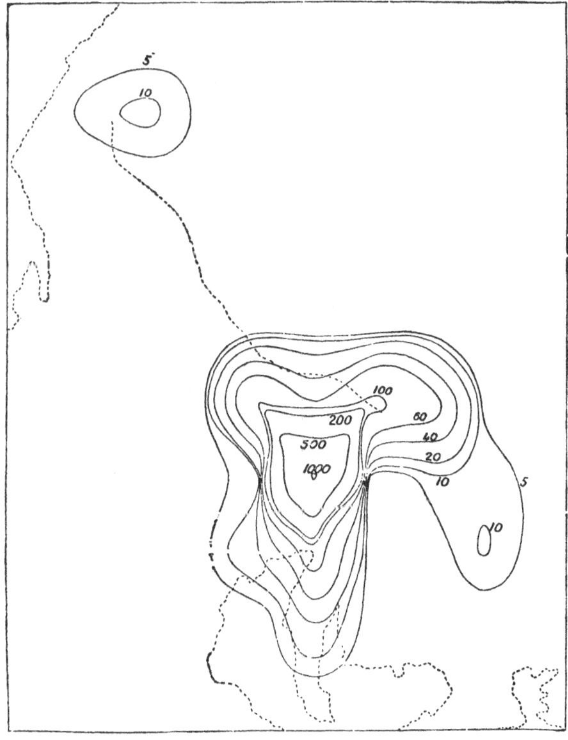

Fig. 86. Distribution of after-shocks of the Mino-Owari earthquake of 1891 (Nov.–Dec. 1891).

The fluctuating, but on the whole increasing, proportion of feeble shocks registered at Gifu is represented by the curve in Fig. 85, which shows the varying percentage in successive months of the feeble shocks with regard to the total number of shocks and earth-sounds.

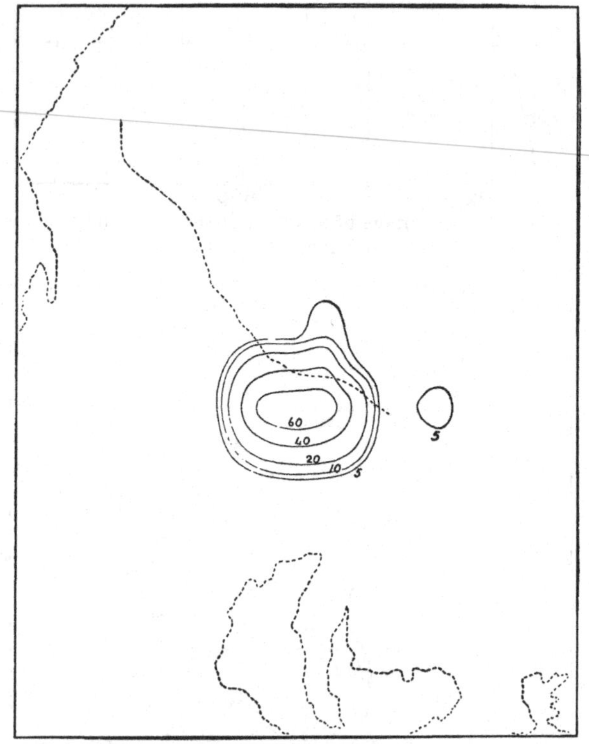

Fig. 87. Distribution of after-shocks of the Mino-Owari earthquake of 1891 (July–Aug. 1892).

215. Focal Migration of After-Shocks. With the lapse of time, after-shocks not only decline in frequency, but also vary in the position of their origin. As in the case of fore-shocks, our knowledge is chiefly based on the after-shocks of the Mino-Owari earthquake of Oct. 28, 1891. The maps in Figs. 86 and 87 are constructed in the same manner as the map showing the distribution of the fore-shocks (Fig. 83).

The first map (Fig. 86) shows the distribution of the after-shock epicentres in November and December 1891, the two months immediately following the great earthquake. The second (Fig. 87) shows the distribution eight months later, in July and August 1892. A comparison of the maps shows at once the decline in frequency; in Fig. 86 there is a curve corresponding to 1000 epicentres, in Fig. 87 the curve of highest order is one corresponding to 60 epicentres. In November and December 1891, after-shocks occurred at the northern end of the fault-scarp, but chiefly in the central and southern regions*. The epicentres cluster along the continuation of the fault-scarp towards the south. The greatest activity, however, is concentrated along the secondary fault, of the existence of which evidence has been already given (sect. 209). Fig. 87 shows that, eight months later, seismic activity had abandoned the terminal regions of the fault. A few epicentres remain near the southern end of the fault-scarp; the larger number are grouped around the more central portion of the scarp, perhaps also along the secondary fault, but in a more northerly portion of it than in the months following the principal earthquake.

The distribution of after-shock epicentres thus points to the existence of a nearly central region of great activity with regions of minor activity near or surrounding the ends of the faults. The activity of these terminal districts was not only less marked, it was also of shorter duration, than that of the central region. At the northern end of the main fault and at the south-eastern end of its continuation, all activity had practically ceased before April 1892. In the region surrounding the southern end of the fault-scarp it lasted until about the close of the same year. A similar withdrawal took place from the southern end of the secondary fault, only two epicentres lying in that district after March 1892. Thus, the distribution of after-shock epicentres of the Mino-Owari earthquake is marked by decrease in the area of activity, and by its gradual but oscillating withdrawal to a more or less central district†.

216. After-Shocks and the Nature of the Principal Earthquake.

* The break between the northern and central groups of curves (Fig. 86) may be apparent rather than real, for the country in this part is mountainous.

† Davison, pp. 5–14. Other diagrams showing the distribution of after-shock epicentres are given on pp. 5–7 of this paper.

The relations between the frequency of after-shocks and the nature of the principal earthquake cannot yet be considered in detail. Two points, however, are worthy of notice.

(i) There seems to be a marked difference in the number of after-shocks which attend twin and other earthquakes. In Great Britain, the earthquakes are invariably either simple or twin. During the 21 years 1889–1909, the three simple earthquakes of Inverness in 1890 and 1901 and Carnarvon in 1903 were followed by at least 33 after-shocks, while 7 twin earthquakes (those of Pembroke in 1892 and 1893, Hereford in 1896, Derby in 1903 and 1904, Doncaster in 1905 and Swansea in 1906) were followed by 18, the ratio per earthquake being 11 to 2·57 or more than 4 to 1. The same relation seems to govern the stronger twin earthquakes of other lands. The Charleston earthquake of 1886 was attended by comparatively few after-shocks; the Riviera earthquake of 1887 by 673 after-shocks from Feb. 23 to Dec. 31, 1887. The Messina earthquake of 1908, with its 949 after-shocks in little more than a year, seems to form an exception to the relation indicated. In this case, however, the focus was evidently close to the surface and changes of elevation were measured along the coasts of the Straits of Messina (sect. 89).

(ii) There is an evident relation between the number of after-shocks of an earthquake and the height of the fault-scarp. In the Mino-Owari earthquake of 1891 and the Assam earthquake of 1897, the fault-scarps were in places of considerable altitude (sect. 81) and the numbers of after-shocks were unusually large. The great Concepcion earthquake of 1835 was probably caused by a displacement along a submarine fault running in a direction parallel to the neighbouring coast-mountains. That it was accompanied by the formation of a fault-scarp is clear from the occurrence of the sea-wave which afterwards swept over the adjoining shores. The coast was also raised by several feet, at one point by not less than 10 feet. The earthquake was followed by hundreds of after-shocks, some of considerable violence, proceeding apparently from the same origin. At the same time, the coast subsided, for, after an interval of some weeks, it stood at a lower level than it did immediately after the principal earthquake.

A large number of after-shocks of the Assam earthquake of

1897 were recorded at Maophlang, where an interesting observation was made. A straight piece of wood was nailed to a stout post so that its upper edge pointed exactly to the crest of a ridge about a mile and a half to the west. Six months later, this edge pointed some way down the slope of the ridge, being apparently tilted through an angle of one degree. The change might be due to a displacement of the post, of which, however, there was no evidence. It probably implies that crustal displacements continued long after the great earthquake, and that they were, in part at any rate, due to the movements which caused the stronger after-shocks.

Lastly, the number of after-shocks of the Californian earthquake of 1906 was extraordinarily small, if we consider the vast length of the fracture and the volume of the displaced crust (sects. 80, 84). During the first 14 months, the total number of recorded shocks in no place exceeded 153. Nor, within the central area, was there the incessant quivering which is so common a feature in the early days after a great earthquake. Indeed, as regards the number of its after-shocks, the Californian earthquake was inferior to a disturbance so slight comparatively as the Comrie earthquake of 1839. In the Californian earthquake, however, the displacement, great as it was, was mainly horizontal. A perceptible fault-scarp was confined to the northern half of the epicentral area, and it was in this district that after-shocks were more frequent than elsewhere. The paucity of Californian after-shocks was no doubt due to the small range of the vertical displacement *.

Sympathetic Shocks

217. Sympathetic Shocks of the Mino-Owari Earthquake. The stresses to which the crust is subjected before and after a great earthquake are not confined to the region of the fault alone. In the whole surrounding country, they must be different after a great earthquake from what they were before. They may be increased or decreased by the displacement which produced the earthquake, and the result may be either an increase or decrease in the seismic activity of the neighbouring regions. So far as

* C. Davison, *Geol. Mag.*, 1910, pp. 417–418; A. Cavasino, *Boll. Soc. Sis. Ital.*, vol. 15, 1911, pp. 142–143; C. Darwin, *Trans. Geol. Soc.*, vol. 5, 1840, pp. 618–619; Oldham, pp. 157–158; Lawson, vol. 1, pp. 410–433.

regards the districts which surround that in which the Mino-Owari earthquake originated, the result was a marked increase in activity. About 45 miles to the east, and 55 miles to the west, of the great fault-scarp are two other districts in which earthquakes are somewhat frequent. In the eastern district 29 earthquakes, and in the western district 20 earthquakes, originated between Jan. 1, 1885, and Oct. 27, 1891. After the earthquake, from Oct. 28, 1891, to the end of 1892, the numbers which originated in the same districts were 30 and 36, in Nov. 1891 alone 7 and 8. Thus, for every earthquake in the eastern district in the period before 1891, 6 were felt in the interval afterwards and 10 in the month of November 1891 alone; for every earthquake in the western district before Oct. 1891, 10 were felt in the interval afterwards, and 16 in November 1891 alone.

The marked increase of seismic activity in these two districts need not, however, be a consequence of the great displacement. The shocks in both central and adjacent districts, it is possible, might result from a general increase of stress over a wide extent of country, and the augmented frequency in the lateral districts could not with justice be regarded as an effect of the former, for they might both be effects of the same widely prevailing cause. But that the connexion is one of real dependence is probable for two reasons. (i) Crustal distortions of the kind and magnitude of those which took place in the Neo Valley could not be effected without a very considerable change of stress in all the surrounding country. (ii) An increase of stress cannot determine the occurrence of an earthquake unless it be sufficient to overcome the resistance to effective displacement. Now, it is unlikely that the gradual increase of stress should be so nearly proportioned everywhere to the prevailing conditions of resistance as to give rise to a marked and practically simultaneous change in seismic activity over a large area; whereas the sudden occurrence of a strong earthquake might alter the surrounding conditions with comparative rapidity, and induce a state of seismic excitement in the neighbourhood. The rapid and simultaneous increase in earthquake-frequency in the two subsidiary districts, distant though they be from one another by 100 miles, seems strongly in favour of this interpretation*.

* C. Davison, *Geol. Mag.*, 1897, pp. 23–27.

CHAPTER XIII

VOLCANIC EARTHQUAKES

Relations between Tectonic and Volcanic Earthquakes

218. Earthquakes have been divided into two main classes (sect. 8)—tectonic and volcanic earthquakes. Tectonic earthquakes are due as a rule to the displacements which effect the growth of faults. Volcanic earthquakes may be defined as those which are caused by the operations which result or tend to result in a volcanic eruption or are due to displacements, by whatever cause they may be produced, along fractures of the volcanic mass, whether the volcano itself be active, dormant or extinct. Thus, volcanic earthquakes are of two kinds: (i) those which are purely volcanic in their origin, and (ii) those which are of tectonic origin in so far as they are due to the growth of faults, but of volcanic origin in that the slips are precipitated by present or past volcanic operations.

219. General Independence of Tectonic Earthquakes and Volcanic Eruptions. A small scale seismic map of the world, such as that of Mallet, conveys the impression that active volcanoes are situated as a rule in regions in which earthquakes are numerous and strong. More detailed maps lead to a different conclusion. For instance, Milne's seismic map of Japan, in which epicentres alone are indicated (sect. 173), shows that "the central portions of Japan, which are the mountainous districts where active volcanoes are numerous, are singularly free from earthquakes." This feature is brought out still more clearly by the map of Japan in Fig. 70. The volcanoes, most of which are active, are represented by black dots, and these, it will be seen, are almost entirely absent from the more darkly shaded areas. Excluding those in the smaller outlying islands, there are 88 volcanoes in Japan. Some of these lie exactly on the border of two of the rectangles into which Milne divides the country (sect. 172), and there are thus 96 rectangles in which volcanoes are entirely or partly situated, while there are 1476 rectangles on land without volcanoes. Now, of the Japanese earthquakes of the years 1885–1892, the average number of epicentres in each non-volcanic rectangle is 3·57, and in each volcanic rectangle 0·73; so that the number of epicentres in a non-volcanic rectangle is very nearly five times as great as in a volcanic rectangle.

Somewhat similar relations govern the occurrence of earthquakes and volcanic eruptions with regard to time. If long intervals of time, one or two centuries, be considered, it appears, as Mercalli has shown, that the intervals, 1632–1737, 1750–1849, in which earthquakes were numerous and strong were also those in which eruptions of Vesuvius were frequent; while the intervals, 1303–1499, 1503–1631, in which earthquakes were of less consequence were also those during which Vesuvius was seldom in eruption. If short intervals of time be considered, this general coincidence, as will be seen later, disappears, and the earthquakes as a rule precede or follow the eruptions and only rarely accompany them*.

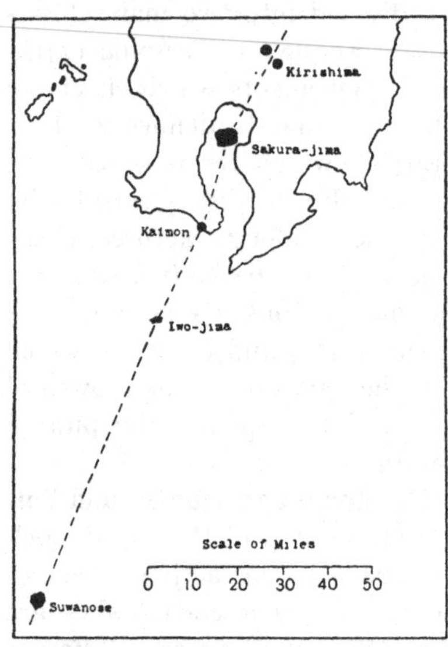

Fig. 88. Volcanic chain of south Japan.

220. A remarkable exception to the last statement, in which volcanic eruptions are occasionally accompanied by true tectonic earthquakes, remains to be noticed. For instance, on Jan. 12, 1914, a few hours after the great eruption of the Sakura-jima (south Japan) began, there was an earthquake strong enough to damage houses within a few miles of the volcano and to be

* R. Mallet, *Rep. Brit. Ass.*, 1858, plate 12; J. Milne, *Seis. Journ.*, vol. 4, 1895, p. xv; C. Davison, *Geogr. Journ.*, vol. 10, 1897, p. 534; G. Mercalli, *Vulcani e Fenomeni Vulcanici in Italia*, 1883, pp. 357–359; J. Milne and H. H. Turner, *Rep. Brit. Ass.*, 1913, pp. 65–67.

recorded in European observatories. One month later, on Feb. 13, a similar earthquake took place during the eruption of the Iwo-jima, a volcano belonging to the same chain as the Sakura-jima. The course of this chain is shown in Fig. 88. On Nov. 18, 1913, the Kirishima-yama broke out in strong eruption; the Sakura-jima followed about two months later; and, one month after that, and still farther to the south, the Iwo-jima. When three volcanoes, situated as these are and all of infrequent activity, burst into eruption so nearly together, and, when two of the eruptions are accompanied by strong and deeply-seated earthquakes, it is difficult not to regard both phenomena as different manifestations of a common cause, namely, the gradually growing stresses along the whole volcanic chain. But there is no reason for supposing that the earthquakes result from the volcanic operations. They should therefore be considered as tectonic, and not as volcanic, earthquakes*.

Earthquakes of Active Volcanoes

221. Etnean Earthquakes. Earthquakes occur on all sides of Etna, but, for some time, they have been specially frequent and violent on its south-eastern flank. Brief descriptions will now be given of a few typical earthquakes.

A great eruption on the east-north-east flank of Etna began on Sep. 10, 1911, and lasted only 13 days. Three weeks later, on Oct. 15, occurred the destructive earthquake of Fondo Macchia, which was preceded by at least ten fore-shocks and followed by five after-shocks, the whole series lasting from Sep. 30 to Nov. 9. The isoseismal lines of the principal earthquake, corresponding to intensities 10, 8, 6, 4, 2 of the Mercalli scale are shown in Fig. 89. The meizoseismal area (bounded by the isoseismal 8) is a slightly sinuous band, 4 miles long, about $\frac{1}{3}$ of a mile wide, and $1\frac{1}{4}$ sq. miles in area. Notwithstanding its great intensity within this band, the shock was not felt at places 6 miles to the west, and the area within the isoseismal 4 was not more than 70 sq. miles.

The dotted line on the same map represents the boundary of

* G. P. Scrope, *Considerations on Volcanos* (1825), p. 155; F. Omori, *Bull. Eq. Inv. Com.*, vol. 8, 1914, pp. 23–24; *Nature*, vol. 92, 1914, pp. 716–717.

the meizoseismal area of the Fondo Macchia earthquake of July 19, 1865, also a narrow band, 5 miles long, $1\frac{1}{4}$ miles wide, and containing about 5 sq. miles. As the shock was felt at no place more than 12 miles from the epicentre, its disturbed area must have been less than 113 sq. miles. It was followed by a number of after-shocks, the last of which occurred on Aug. 23

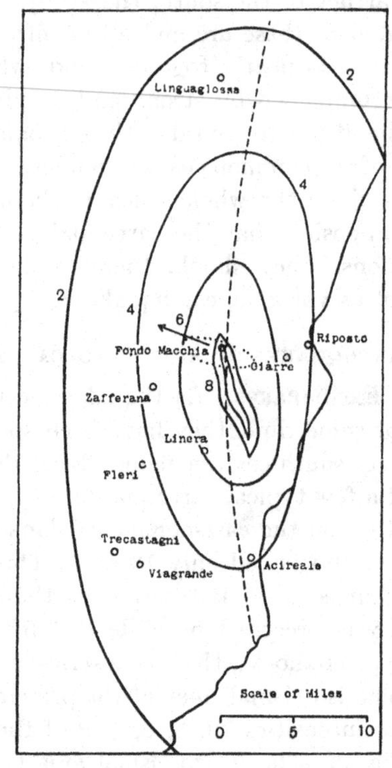

Fig. 89. Map of the Fondo Macchia earthquakes of 1865 and 1911.

This earthquake also succeeded a great eruption of Etna, which began on Jan. 30, 1865, and lasted for nearly 12 weeks. The direction from Fondo Macchia of the central crater of Etna is represented by the arrow in Fig. 89.

One of the most remarkable series of Etnean earthquakes occurred in the neighbourhood of Linera on May 8, 1914. The total number of sensible shocks was 55, 21 being fore-shocks from Apr. 28 to May 7, and 33 after-shocks from May 8 to

June 4. The curves in Fig. 90 represent the boundaries of the areas within which houses were damaged by the more important shocks, *A* and *B* of the double earthquake of May 7 at 6.35 p.m., *C* of another earthquake on the same day at 10 p.m., the fracture within it being probably a continuation of that within the area *B*, *D* and *E* of the principal earthquake of May 8, and *F* that of the after-shock of May 26. Within the area bounded by the

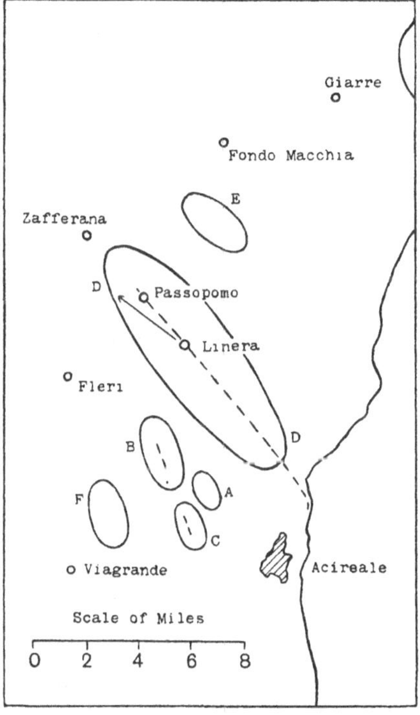

Fig. 90. Map of the Linera (Etna) earthquakes of May 8, 1914.

curve *D*, which is about 4¼ miles long and 1¼ miles wide, not only were houses completely razed to the ground, but the ground itself was crushed. Along the axis of this zone, there runs a slightly sinuous fracture, from Passopomo through Linera to the sea, its course being represented by the broken line on the map. In almost every part of it, there is a change of level, of an inch or two only in some places, in others of 15 or 16 inches, and in one, near the sea-coast, of more than 3 feet, the ground on the south-west side being left at the higher level. The direction of the central crater from Linera is represented by the arrow in the map*.

Thus, the longer axes of the isoseismal lines of the earthquakes of 1865 and 1914 are directed towards the central crater, while those of the earthquake of 1911 are nearly at right angles to that direction. These earthquakes are types of two classes of

Fig. 91. Seismic districts of Etna.

Etnean earthquakes, the majority of which are probably connected with radial fractures of the volcano, and others with peripheral fractures†.

222. Most Etnean earthquakes, as in the above examples,

* During the interval covered by this series of earthquakes, there was a marked increase in the activity of Etna, though the shocks were not coincident with volcanic explosions.

† A. Riccò, *Boll. Soc. Sis. Ital.*, vol. 16, 1912, pp. 9–32; M. Baratta, *Boll. Soc. Geogr. Ital.*, Oct. 1894; G. Platania, *Pubbl. dell' Ist. di Geogr. Fis. e Vulcan. della R. Univ. di Catania*, No. 5, 1915.

disturb very small areas, and the villages in which houses are damaged are in all such cases close to the epicentres. Baratta has shown that Etnean earthquakes are especially frequent in 12 zones, which he names after neighbouring towns or villages. These are given in the accompanying map (Fig. 91), the dotted lines joining several places (such as Adernò, Bronte and Maletto) indicating that their environs form a single zone.

Though the zones on the south-eastern flank of Etna are at present those which are most frequently in action, earthquakes are by no means confined to them, and are indeed subject to frequent transferences from one zone to another. For instance, in one year only (1903), 8 of the 12 zones were in action*. In any one zone, the epicentres are not absolutely fixed, though, time after time, the same village in a zone (*e.g.* Nicolosi or Fondo Macchia) may be damaged or destroyed†.

223. Japanese Volcanoes. Some interesting observations with a horizontal-pendulum seismograph have been made by Omori on the minute tremors which accompany volcanic eruptions in Japan.

The Asama-yama (central Japan), which rises to a height of 8140 feet above the sea, was subject to a series of strong explosions during the years 1908–1914. The observations were made in the summer months at a station, 6306 feet high, on the south-western flank. The seismograms are divided by Omori into two classes, (i) those due to earthquakes which were not accompanied by any outburst of the volcano, and (ii) those due to earthquakes which were invariably coincident with explosions. The former consisted only of minute quick vibrations; the latter began with slow movements on which, after a few seconds, quick vibrations were superposed. The earthquakes without explosions were distinctly the stronger—of the 1485 shocks recorded from 1911 to 1916, 21 per cent. were sensible; while of the 8847 earthquakes with explosions, only 0·3 per cent. were

* Namely, the zone of Linguaglossa on Apr. 7, of S. Maria di Licodia (probably) on Apr. 13 and July 30; of Paternò on Apr. 19; of Belpasso on Mar. 24 and July 21; of Nicolosi on Mar. 8 and probably on Aug. 6 (4 shocks); of Trecastagni on May 26 and June 1–16 (23 shocks); of S. Venerina, etc., on Mar. 11 and Nov. 20; and of Giarre-Riposto on Jan. 30.

† M. Baratta, *I Terremoti d' Italia*, 1901, pp. 829–833; S. Arcidiacono, *Bull. dell' Accad. Gioenia di Sci. Nat., in Catania*, Fasc. 79, 1903.

sensible. Again, the two types of earthquakes alternate in frequency, the maxima of one type occurring at about the same time as the minima of the other*.

224. The last eruption of the Usu-san (north Japan) occurred in 1910, and consisted of explosions in a series of craterlets arranged along a peripheral fracture of the volcano. Seismometric observations were made at two places, one (West Kohan) close to the east end of the line of craterlets, the other (Nishi-Monbets) 5 miles from the crater. At the former station, but not at the latter, series of well-defined minute quick vibrations, called micro-tremors by Omori, were recorded. The mean range of motion was always less than one-tenth of a millimetre, and the principal periods (·53, 1·08, 1·59 and 2·14 seconds) were practically identical with those of earthquake-vibrations registered at Nishi-Monbets. Omori therefore concludes that micro-tremors are true earth-vibrations, but so weak that they cannot be recorded more than a few miles from the origin. Violent explosions in the craterlets were usually accompanied, and sometimes preceded, by micro-tremors†.

EARTHQUAKES OF DORMANT AND EXTINCT VOLCANOES

225. Ischian Earthquakes. Ischia is a small island 6 miles from the west coast of Italy and about 20 miles from Naples. The central crater of M. Epomeo has long been extinct, but an eruption occurred from a lateral cone in the year 1302. The volcano may therefore be regarded as dormant. From this year until 1796, the island was practically free from earthquakes. A series of earthquakes then began, strong enough to damage houses in Casamicciola, followed by others in 1828, 1841, 1867, 1881 and 1883, by the last of which Casamicciola was ruined. In the map (Fig. 92), the dotted lines represent the boundaries of the existing portions of Epomeo, and the continuous lines the boundaries of the area within which buildings were seriously damaged by the earthquakes of 1796, 1828, 1881 and 1883. The broken line shows the position of the radial fracture with which the earthquakes were connected, and it was along this line, in the neighbourhood of Casamicciola, that the chief damage in 1881 and 1883 was concentrated.

* F. Omori, *Bull. Eq. Inv. Com.*, vol. 7, 1917, pp. ii–iii.

† *Ibid.*, vol. 5, 1911, pp. 31–38.

In 1881, as Johnston-Lavis has shown, the area of complete destruction included only half a sq. mile, the area of serious damage about 2 sq. miles, and that of partial damage about 5 sq. miles, while the shock was just felt on the Italian coast, which lies about 10 miles from Casamicciola. In the much stronger earthquake of 1883, the areas of complete destruction, serious damage and partial damage were respectively 3, 11 and 30 sq. miles, while the shock was felt by a few persons in Naples. In both earthquakes, the depth of the focus, as estimated from the inclination of fissures in buildings, was found to be about

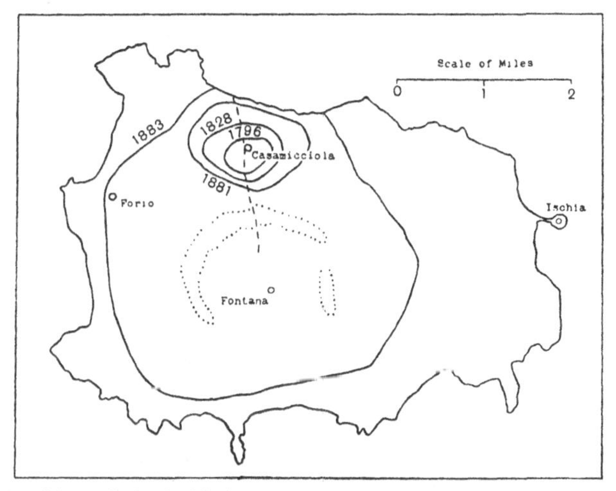

Fig. 92. Map of the Ischian earthquakes of 1796, 1828, 1881 and 1883.

one-third of a mile. The earthquake of 1883 was remarkable for the entire absence of preliminary sound and tremor and for its great initial strength, most of the ruin being caused during the first few seconds. Both earthquakes were followed by a few slight after-shocks*.

226. Alban Hills Earthquakes. The Alban Hills form a group of extinct volcanoes, the principal hill, M. Cavo, being about 15 miles south-east of Rome. The earthquakes are less frequent and less destructive than those of the Etnean region, but in other respects they resemble them closely. Baratta defines nine

* G. Mercalli, *L' isola d' Ischia ed il terremoto del* 28 *luglio* 1883 (Milan, 1884); H. J. Johnston-Lavis, *Monograph of the Earthquakes of Ischia* (1885).

seismic zones within the area of the hills, and, while two or three zones are at present more active than the others, there is continual migration from one zone to another, the longer axes of the isoseismal lines being occasionally peripheral but more often radial. One of the most recent earthquakes in the district is that of Albano on Feb. 21, 1906. It was preceded by six very slight fore-shocks and followed by six very slight after-shocks, the whole series being included between Feb. 20 and 23. Near the epicentre, the intensity of the shock was 5–6 (Mercalli scale), and the close grouping of the isoseismal lines indicates a rapid decline outwards in the strength of the shock. The mean duration of the shock was nearly 4 seconds*.

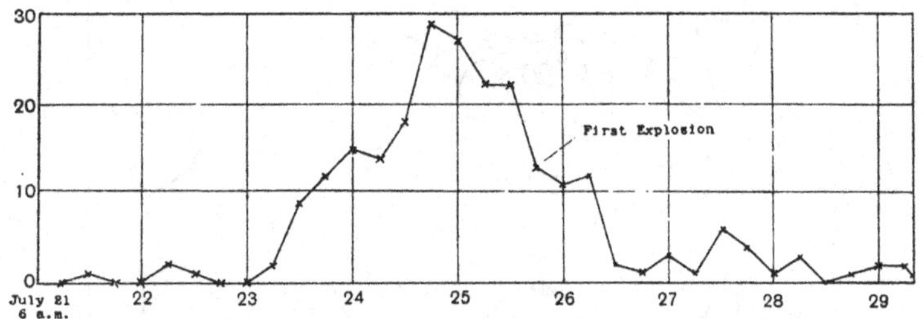

Fig. 93. Variation in earthquake-frequency before and after the eruption of the Usu-san (Japan) in 1910.

Characteristics of Volcanic Earthquakes

227. (i) **Frequency before an Eruption.** Volcanic eruptions are usually preceded by a large number of fore-shocks, one or two of which may attain considerable strength. The last eruption of the Usu-san (north Japan) began at 10 p.m. on July 25, 1910. The first shock was felt at Nishi-Monbets (5 miles from the volcano) on July 21; on the next four days, the numbers felt were 25, 110, 351 and 165. Once the eruption had begun, the shocks became less frequent. At Sapporo, about 44 miles from the volcano, the earthquakes were registered by a seismograph, the numbers being 1 on July 21, 3 on July 22, 23 on July 23, 76 on July 24, and 84 on July 25; after the eruption had begun, the numbers were 26 on July 26, 15 on July 27, 5 on July 28, 6 on July 29, and 1 on July 30. The variation in six-hourly

* M. Baratta, *I Terremoti d' Italia*, pp. 778–780; G. Agamennone, *Boll. Soc. Sis. Ital.*, vol. 21, 1918, pp. 47–101.

frequency is represented in Fig. 93, from which it is clear that the shocks had begun to decline in frequency a few hours before the first explosion occurred. The last great eruption of the Sakura-jima (south Japan) began at 10 a.m. on Jan. 12, 1914. At Kagoshima (about 6 miles from the centre of the volcano), 418 earthquakes were recorded during the 31 hours preceding the first eruption, the average hourly frequency being 4·1 from 3 to 11 a.m. on Jan. 11, 12·4 from 11 a.m. to 8 p.m., and 19·5 from 8 p.m. on Jan. 11 to 10 a.m. on Jan. 12. After the explosion at 10 a.m., the successive hourly numbers were 17, 11, 6, 3, 5, 2, 2 and 2, the seismograph being then injured by the strong earthquake referred to above (sect. 220)*.

(ii) **Small Disturbed Area.** No feature of volcanic earthquakes is so marked and none so significant, as the smallness of the disturbed area, considering the great intensity of the shock at the epicentre. We may have an earthquake, like that of Nicolosi in 1901, destroying houses within a minute meizoseismal area and yet imperceptible at a distance of more than 4 miles, or one like that of Fondo Macchia in 1865, causing utter ruin over an area of 5 sq. miles and yet not felt outside an area of more than 113 sq. miles, or, again, one like that of Ischia in 1883, levelling every building within an area of 3 sq. miles, and only just perceptible at a distance of 20 miles from the epicentre.

(iii) **Sharpness and Brevity of the Shock.** To the unaided senses, the shock of a volcanic earthquake is remarkable for its sudden onset, its sharpness and brevity. Preliminary sound and tremor are either absent or else of duration so brief as to be almost imperceptible. Even animals show no sign of disquietude, the shock is most violent at its onset, so that the ruin of houses is almost instantaneous. The duration of the shock is usually short, being often less than 5 seconds.

(iv) **Small Seismic Focus.** The length of the focus is inconsiderable. In the strongest of Etnean earthquakes, it seldom exceeds 3 or 4 miles; in the Ischian earthquakes of 1881 and 1883, it was less than a mile; in the Albano earthquake of 1906, it was about 3 miles. This conclusion is also supported by the brief duration of the shock.

* F. Omori, *Bull. Eq. Inv. Com.*, vol. 5, 1911, pp. 8–17; vol. 8, 1914, pp. 9–14, 22–27.

(v) **Accessory Shocks limited in Time and Space.** Some, but not all, volcanic earthquakes are preceded and followed by accessory shocks. In this, they resemble tectonic earthquakes. But the after-shocks of volcanic earthquakes are distinguished by the short period of their action. For instance, the after-shocks of the Albano earthquake of 1906 lasted for 3 days, in the Ischian earthquake of 1883 for 7 days, in the Fleri earthquake of 1914 for about 3 weeks, in the Linera earthquake of 1914 for nearly 4 weeks, and in the Fondo Macchia earthquake of 1865 for 5 weeks, Fleri, Linera and Fondo Macchia being on the south-eastern flank of Etna. Again, the after-shocks of volcanic earthquakes are practically confined to the epicentral area. They point to little, if any, tendency towards an extension of the original focus.

(vi) **Stability of Epicentre.** In a volcanic system, such as that of Etna or the Alban Hills, earthquakes occur in many different zones and seismic activity is subject to frequent and sudden migrations from one zone to another. Nevertheless, in any given zone, there is often a certain fixity in the epicentres of successive earthquakes. In the Alban Hills, villages such as Frascati, Albano or Ariccia are shaken, while the surrounding country is almost undisturbed. Since the beginning of the 14th century, every earthquake of any consequence in Ischia has originated in the same district close to Casamicciola. In the Etnean zones, time after time, the same places, such as Nicolosi, Fondo Macchia, etc., are damaged or destroyed.

228. The feature in which volcanic earthquakes differ most widely from tectonic earthquakes is the great intensity of the shock near the centre of a very small disturbed area. That this rapid decline in intensity is due to the shallowness of their foci is clear from sect. 139; and this inference is supported by two other lines of evidence. In the Ischian earthquakes of 1881 and 1883, the depth of the focus was estimated to be one-third of a mile (sect. 133). Again, the duration of the preliminary tremor of an earthquake increases with the distance from the focus, and the brevity or practical absence of any such tremor in volcanic earthquakes shows that the foci must be very close to the surface.

229. We may thus conclude: (i) that the foci of volcanic earthquakes are very shallow; (ii) that the foci are usually small and not often more than 4 or 5 miles in length; (iii) that the accessory shocks are practically confined to the original focus; and (iv) that, while most volcanic earthquakes originate along radial fractures of the mountain, some—and those not the least important—originate along peripheral fractures.

Origin of Volcanic Earthquakes

230. The earthquakes which occur near or below active and dormant volcanoes are naturally attributed to the proceses which tend to result in a volcanic eruption, the causes appealed to being: (i) the formation of new fractures, or the re-opening or extension of old fractures, in the mountain-mass; (ii) explosions, due to any cause, within the volcano; (iii) the sudden injection of lava into fractures or cavities in the mass of the volcano; and (iv) the relative displacement of the rock-masses adjoining a fracture of the volcano.

In active volcanoes, there can be no doubt that the formation or extension of fractures and (especially) explosions give rise to a large number of very slight earthquakes. But, as we know from the evidence of the Japanese seismograms, the earthquakes accompanying explosions are weak and the great majority imperceptible to the unaided senses. Nor is there any reason for supposing that the process of fracturing would cause shocks of much strength, and, in any case, the frequent repetition of shocks in the same region would be difficult to explain on this theory. The sudden injection of lava into fractures or cavities of the mountain would be a more efficient cause of strong earthquakes, especially beneath the flanks of a dormant volcano like M. Epomeo. The theory may be held to account satisfactorily for many of the phenomena of volcanic earthquakes; it is hardly applicable to the earthquakes of an extinct volcanic district.

On the whole, it is probable that the more important volcanic earthquakes are due to the relative displacement of the rock-masses adjoining a fracture of the volcano. We know, from the evidence of the Linera earthquake of 1914, that such displacements do occur; it is clear that the resulting friction, depending as it does on the weight of the mass displaced, must be capable of producing such strong earthquakes as those which visit the flanks of Etna and Epomeo.

231. In some of the mining districts of Great Britain, there are occasionally earth-shakes which, though of much less strength than the Etnean earthquakes, resemble them closely in their nature, and probably also in their origin. In Fig. 94 are shown the isoseismal lines of an earth-shake which occurred at Pendleton (near Manchester) on Nov. 25, 1905. The shock was of intensity 7 (Rossi-Forel scale) near the centre of a disturbed

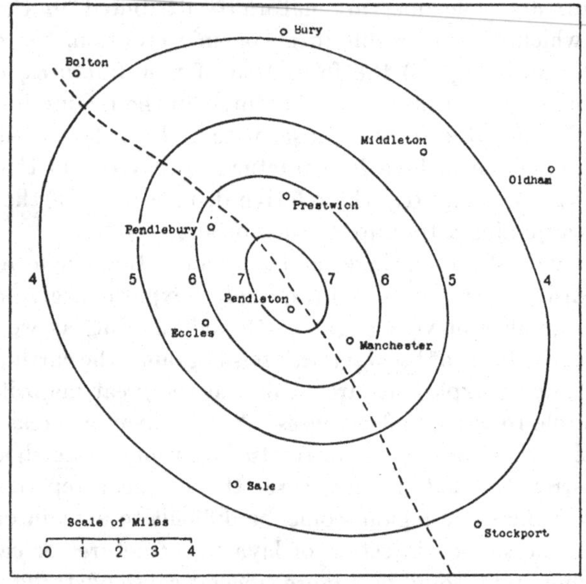

Fig. 94. Map of the Pendleton earth-shake of Nov. 25, 1905.

area containing 144 sq. miles. As the average disturbed area of British earthquakes of the same intensity is 24,500 sq. miles, it is evident that the earth-shake originated at a very slight depth. The mean direction of the longer axes of the isoseismal lines is N. 37° W., that of the Irwell Valley fault near the epicentre (represented by the broken line in Fig. 94) is N. 34° W. It therefore seems probable that the earth-shake was due to a slip along this fault, caused either by the pumping of water from the mine or by the removal of the coal up to the face of the fault. If so, the earth-shake was of natural origin in so far as it was due to the growth of the fault, but of artificial origin in that the slip was precipitated by mining operations*.

232. An active volcano is traversed by many fractures, the majority of which are radial, but a few are peripheral. If one or both of the masses adjoining a fracture were to be deprived of support, the resulting displacement would give rise to an earthquake, not to be distinguished, except by its scale, from a true tectonic earthquake. The support might be withdrawn either by underground movements of the magma†, or, less frequently, by the cooling of a mass of lava or heated rock below, by the former in active volcanoes, and by the latter especially in dormant and extinct volcanoes. This theory, it will be seen, accounts for all the known phenomena of volcanic earthquakes—for their close connexion in space and time with many eruptions, the shallowness and small size of the foci, the frequent repetition in the same region, the intensity and brevity of the shocks, and the occurrence of series of fore-shocks and after-shocks of brief duration and limited zone of displacement.

* C. Davison, *Geol. Mag.*, 1905, pp. 219–223; 1906, pp. 171–176.

† G. Platania, *Pubbl. dell' Ist. di Geogr. Fis. e Vulcan. della R. Univ. di Catania*, No. 5, 1915, pp. 1–2, 41.

CHAPTER XIV

ORIGIN OF TECTONIC EARTHQUAKES

EARTHQUAKES AND THE GROWTH OF FAULTS

233. An earthquake has been defined as the result of any sudden displacement within the earth's crust (sect. 1). Of the displacements known or inferred, the following have been suggested as possible causes of earthquakes:

(i) The fall of rock in underground channels;

(ii) The operations which result or tend to result in volcanic eruptions; connected with which are

(iii) Explosions of suddenly generated steam when water filtering through the outer crust reaches the highly-heated rock below;

(iv) The fracturing of the solid crust; and

(v) The intermittent growth of faults, the usual cause being the friction generated by the sudden sliding of one rock-mass against the other, but complicated sometimes, when the displacement extends to the surface, by the movement of the mass as a whole*.

To the first of these causes we may perhaps attribute some slight local shocks. The earthquakes connected with volcanic eruptions have been considered in the preceding chapter. Often destructive within a limited area, they are seldom felt more than a few miles from the origin, and the greatest of all earthquakes occur in regions which are far removed from present or

* The following references may be given to memoirs in which the above theories are considered at greater length than is here possible:

1. Davison, C. (1). Twin-earthquakes. *Quart. Journ. Geol. Soc.*, vol. 61, 1905, pp. 18–33.

2. —— (2). *The Origin of Earthquakes* (Cambridge Univ. Press), 1912.

3. Lebour, G. A. On the breccia-gashes of the Durham coast and some recent earth-shakes at Sunderland. *Trans. N. of Eng. Inst. of Min. Eng.*, vol. 33, 1884, pp. 165–174.

4. See, T. J. J. The cause of earthquakes, mountain formation and kindred phenomena connected with the physics of the earth. *Proc. Amer. Phil. Soc.*, vol. 45, 1906, pp. 274–414.

past volcanic action. To account for such earthquakes, the third cause has been invoked. We have no proof, however, of the occurrence of such explosions, nor does the theory provide a complete explanation of the distribution and phenomena of earthquakes.

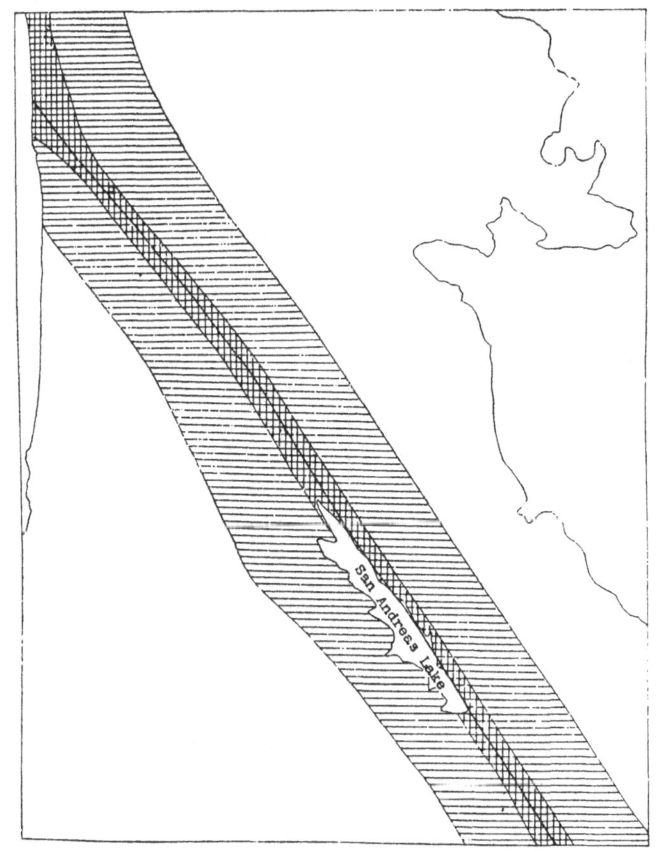

Fig. 95. Portion of the meizoseismal area of the Californian earthquake of Apr. 18, 1906.

The principal reasons for connecting earthquakes with fault-slips are the following: (i) with every step in the growth of a fault, it is evident that an earthquake must occur; (ii) in some great earthquakes, the fault-displacements are manifest; (iii) in all but the weakest earthquakes, the inner isoseismal lines are elongated in form, their longer axes being parallel to the fault-lines of the district; (iv) the number of earthquakes in any region far exceeds the number of the faults; (v) in a series of associated earthquakes, the epicentre migrates to and fro in the direction of the fault; and (vi) owing to variations in the volume and displacement of the rock-mass, fault-slips are capable of producing the weakest tremor as well as the most violent shock.

234. Fault-Displacements. In the somewhat rare cases in which fault-displacements occur (see Chapter V) the connexion between the origin of the earthquake and the growth of the fault is evident. In some cases, as in the Mino-Owari earthquake of 1891 and the Californian earthquake of 1906, it is further shown by the manner in which the inner isoseismal lines cling to the course of the fault. In Fig. 95, for instance, the central continuous line represents a portion of the San Andreas fault (sect. 84) to the south of San Francisco. The darkly-shaded area is that which lies within the isoseismal 10 (Rossi-Forel scale), and the lightly-shaded area that which lies within the isoseismal 9. In no part of its course, which includes the entire land-area of the fault-displacement, does the width of the inner band exceed 2 miles; and, throughout its whole extent, the fault-line runs almost centrally between its boundaries.

That the earthquake in such cases is to be attributed to the growth, rather than to the formation, of the fault, is evident from the fact that in some earthquakes—such as those of Baluchistan in 1892, Alaska in 1899, and California in 1906—the movement has taken place along pre-existing faults.

235. Elongated Forms of Isoseismal Lines. The forms of the isoseismal lines of some British earthquakes are shown in Figs. 28, 30, 50, 51. In these cases, the inner curves are elongated, while the outer curves are nearly circular in form. In the Hereford earthquake of 1896, the dimensions of the innermost isoseismal line are 40 and 23 miles; in the Inverness earthquake of 1901 (Fig. 50) 12 and 7 miles; and in the Derby earthquake of 1903 (Fig. 28), 16½ and 8 miles. In many other British earthquakes, the elongation is equally marked. In one of the Wells earthquakes of 1893, the dimensions are 11½ and 5 miles; in the Exmoor earthquake of 1894, 23 and 12 miles; in one of the Carlisle earthquakes of 1901, 37 and 13 miles; in the Carnarvon earthquake of 1903, 33½ and 15 miles; and in the Swansea earthquake of 1906, 26 and 14 miles. A few examples of other earthquakes may be given. In the Locris earthquake (north-east Greece) of 1894, the dimensions of the innermost isoseismal are 17 and 5 miles; in the Constantinople earthquake of the same year, 109 and 24 miles; and, in the Baluchistan earthquake of 1909, 57 and 8 miles.

Two explanations of the elongation of the isoseismal lines are possible—one that the vibrations are transmitted with less loss of energy in the direction of their longer axes than in the perpendicular direction; the other, that the seismic focus is of considerable length and parallel to the longer axes. If the former explanation were correct, the isoseismal lines of an earthquake should be approximately similar in form, and so also should be those of different earthquakes in the same region. But, as will be seen from the map of the Derby earthquake of 1904 (Fig. 51), the innermost isoseismal line may be circular and the next elongated. In the Doncaster earthquake of 1905, the innermost isoseismal consists of two detached circular portions, while the next is elongated. In the series of Inverness earthquakes of 1901 (Fig. 96), the isoseismal lines of some shocks are elongated and of others circular. Lastly, in the Pembroke earthquakes of 1892, the isoseismal lines of some earthquakes are directed north and south, of others east and west.

Thus, though the forms of isoseismal lines may be, and are, modified by the nature of the surface-rocks (sects. 45, 46), it seems clear that their elongated forms are chiefly due to the existence of seismic foci of considerable length directed parallel to the longer axes. Moreover, since these axes are almost invariably parallel or perpendicular to the principal faults of the epicentral district, and since in many cases they are known to lie on the downthrow side of the faults, it follows that the earthquakes must be associated with the faults, most often with the strike-faults, but not seldom with the transverse faults, of the district.

236. Number of Earthquakes greatly in excess of the Number of Faults. Few earthquakes could result from the formation and extension of a fracture, whereas the subsequent growth of the fault must be the result of innumerable slips. Now, the number of earthquakes felt in a district is far in excess of the number of faults, and our earthquake-records extend over only a few years, rarely over centuries, whereas the formation of many faults has occupied a large part of geological time. We need only refer to the 143 earthquakes noticed at Comrie, in Perthshire, during the last three months of 1839; to the 306 shocks felt at Zante in the year 1896; or to the 3365 earthquakes recorded at Gifu in Japan, from Oct. 29, 1891, to the end of 1893. In such cases, the earthquakes must be due to the growth of existing faults rather than to the formation of new fractures.

237. Migration of the Epicentre along a Fault. Examples of the migration of the epicentre are given elsewhere in this volume —in the case of the Inverness earthquakes of 1901 in sect. 240 and of the Mino-Owari earthquakes of 1891 in sect. 215. One other instance may be given--that of the Carnarvon earthquakes of June 19, 1903, and 1906, connected with the Aber-Dinlle fault. Denoting slips at the north-eastern end, centre and south-western end by the letters *N*, *C*, *S*, the distribution of the different epicentres in time is represented as follows, from left to right—

N, principal focus, *S*, *N*, *N*, *N*, *N*, *C*, *N*, *N*, *C*, *N*, *C*, *C*, *C*—*N*, the last occurring after the lapse of three years.

Now, if the migration were to take place outwards in one or both directions only, this migration might be due to a gradual extension of the fracture; but, since the epicentres retrace their steps, returning to the central region, the corresponding shocks must be referred to fault-slipping rather than to an extension of the fracture.

238. Adequacy of Fault-Growth as a Cause of Earthquakes. It is important to notice that the generation of an earthquake by fault-slipping requires a far less expenditure of energy than that by the formation of new fractures, and this is a matter of consequence seeing that, in all probability, the origin of both must be traced to the slow cooling of the earth. Moreover, in the case of fracturing, the initial disturbance is merely the elastic recoil of rock-particles from the surface of the fracture; whereas, in the case of slipping, the initial disturbance depends on the mass of the displaced crust, increased in some earthquakes by the sudden motion of the crust. As the weight of the mass may vary from that of a few cubic miles to that of several thousand, or even million, cubic miles, and the displacement from a small fraction of an inch—a mere creep—to several or many yards, it is evident that the friction so generated must be capable of producing earthquakes of every degree of strength—from the slight shocks which are occasionally felt in this country to the destructive earthquakes which visit the greater seismic districts, such as those of Calabria, Central Asia, Chili and Japan.

ORIGIN OF SIMPLE EARTHQUAKES

239. Earthquakes have been divided, according to the nature of the shock, into the three classes of simple, twin and complex earthquakes (sect. 34). The great majority of slight and moderately strong earthquakes belong to the first class. In Great Britain, for instance, 95 per cent. of the earthquakes are simple, and the remainder twin, earthquakes, the latter being usually of much the greater intensity. The Inverness earthquakes of 1901 may be taken as typical examples of simple earthquakes, and the Derby earthquakes of 1903 and 1904 of twin earthquakes. Complex earthquakes, of which the Mino-Owari earthquake of 1891, the Assam earthquake of 1897, and the Alaskan earthquakes of 1899, may be regarded as types, will be considered more briefly, as they have already been referred to in Chapter V.

240. Inverness Earthquakes of 1901. The isoseismal lines of the principal earthquake, which occurred on Sep. 18, are represented in Fig. 50; and it has been shown (sect. 129) that the earthquake was probably due to a slip along the great fault which traverses the whole of Scotland in a south-westerly direction past Inverness, and which is responsible for the marked linearity of its surface-features from Tarbat Ness to Loch Linnhe.

Further evidence in support of this connexion is given by the accessory shocks. The boundaries of the disturbed areas of the more important shocks are shown in Fig. 96, and the centres of these areas in Fig. 97. There was first a slight fore-shock (*a*) at 6.4 p.m. on Sep. 16. Then came the principal earthquake (*B*) at 1.24 a.m. on Sep. 18, the focus extending nearly from Inverness to Loch Ness. At about 1.35 a.m., a slight after-shock (*c*) originated near the south-west margin of the principal focus. At 3.56 a.m., the strongest after-shock (*g*) occurred. Its centre was half a mile farther to the north-east, but, as its focus was several miles in length, it must have extended some distance beyond the south-west margin of the principal focus. At 9 a.m., another shock (*h*) occurred, with its

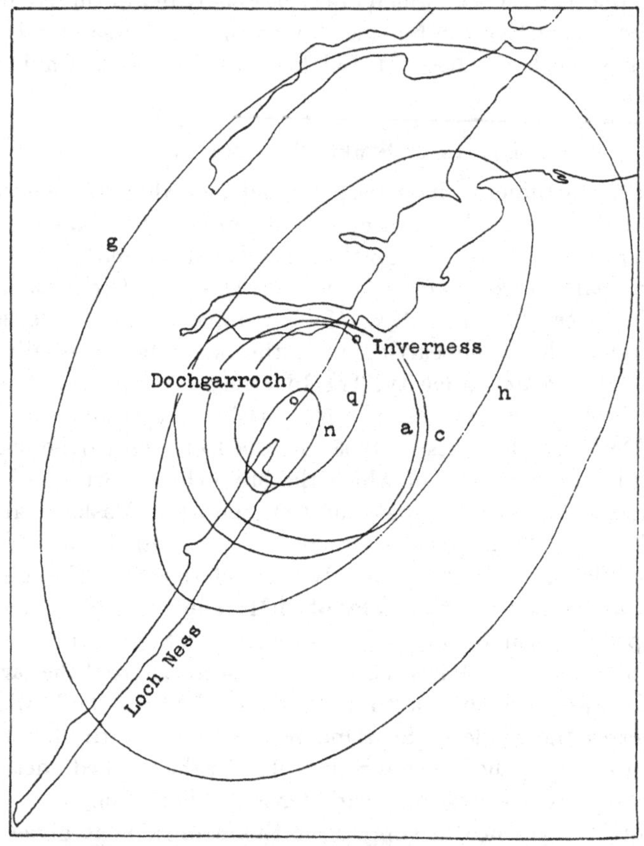

Fig. 96. Map of the principal after-shocks of the Inverness earthquake of 1901.

centre half a mile north-east of the principal centre (*B*), and with a focus slightly overlapping the north-east margin of the principal focus. The next important movement occurred on Sep. 29, with its centre about 1 mile to the south-west of the principal centre. This was followed, on Sep. 30, by one of the strongest after-shocks (*n*), the centre of which lay to the south-west of the principal centre, and the focus of which must have extended 2 or 3 miles beneath Loch Ness. Again, on Oct. 13, occurred the last strong after-shock (*q*), with its focus in the neighbourhood of Dochgarroch. Thus, the foci of the principal earthquake and of all the accessory shocks lie on the downthrow

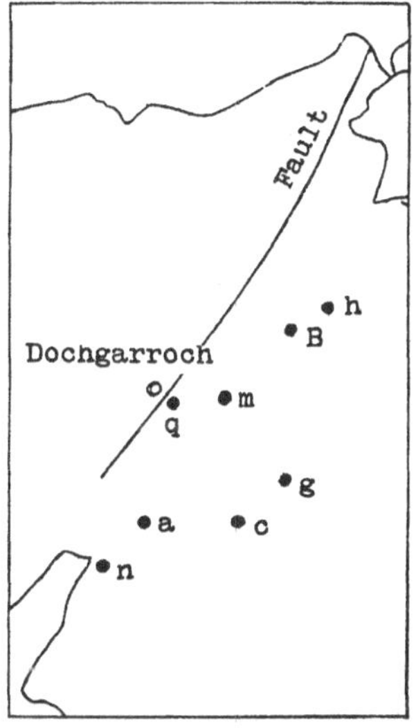

Fig. 97.- Distribution of the centres of the principal after-shocks of the Inverness earthquake of 1901.

side of the fault. It will be noticed also that, in the latter shocks, there is a gradual approach of the centre towards the fault-line, showing that the depth of the corresponding foci gradually decreased with the lapse of time.

241. Origin of Simple Earthquakes. In the case of the Inverness earthquakes, there is no evidence to show on which side of the fault the rock was displaced. If, however, the movements were a continuation of those which have occurred in the past, it is probable that the principal earthquake was produced by a slight but sudden sag of the crust on the south-east side extending from Inverness to near the end of Loch Ness. In the diagram (Fig. 98), the upper line is supposed to represent the surface of the earth from the Moray Firth (north-east of Inverness) to Loch Ness. The straight dotted line is intended to represent a horizontal straight line traced on the south-east side of the fault-surface through the focus of the principal earthquake. Owing to the movement which resulted in the earthquake, this line becomes a curve represented by the continuous line *ACB*. The distance *AB* represents the length of the focus, about 8 miles. The distance between the straight dotted line and the curve at *C* represents the amount of the subsidence. In the diagram, this distance is greatly exaggerated. In reality, it may have been a fraction of an inch.

The first effect of this displacement would be an increase of stress in the terminal regions, *A* and *B*, of the principal focus. If the rock were previously near the point of slipping, the additional stress would be sufficient to cause slips in these regions. The effects of these secondary slips would be new increases of stress still farther outwards and again in the central

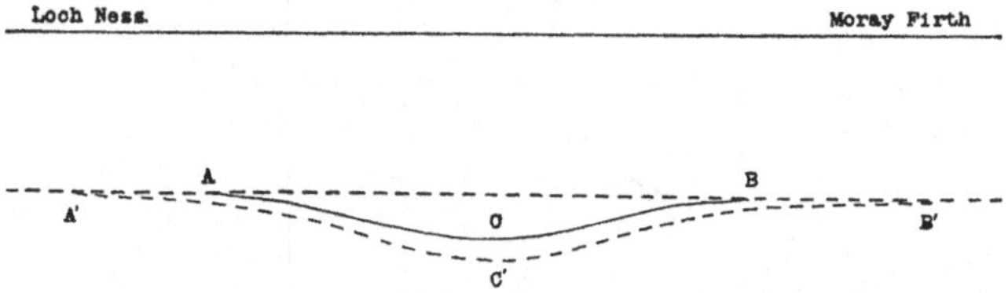

Loch Ness Moray Firth

A B A' C B' C'

Fig. 98. Diagram illustrating the nature of the displacement that causes a simple earthquake.

region below *C*, and this would continue until the additional stresses imposed were no longer able to overcome the resistance to movement. The final form of the straight dotted line *AB* would thus be represented by the curved dotted line *A'C'B'*. It is evident, also, that a downward movement at the level of the focus would increase the stresses on the fault-surface in the central region above. Thus, there would be a tendency for the foci to diminish gradually in depth.

Now, of the six principal after-shocks of the Inverness earthquake, the first and second occurred at and slightly beyond the south-west margin of the principal focus, the third at and beyond the north-east margin, the fourth near the centre and nearer the surface, the fifth beyond the south-west margin and at a still smaller depth, while the last was in the central region and quite close to the surface*.

* C. Davison, *Quart. Journ. Geol. Soc.*, vol. 58, 1902, pp. 377–397.

Origin of Twin Earthquakes

242. Derby Earthquakes of 1903 and 1904. The isoseismal lines of the Derby earthquake of Mar. 24, 1903, are represented in Fig. 28. The two inner isoseismals are elongated in the direction N. 33° E., and they are farther apart on the north-west side. It may be inferred (sect. 129) that the average direction of the fault is N. 33° E., and its hade to the north-west. The fault is probably deep-seated, there being none known in this position.

In the greater portion of the disturbed area, the shock consisted of two distinct parts separated by a brief interval of rest and quiet. At some places, only one shock was felt. Plotting all such places on the map, they are found to lie within a straight narrow band, about 5 miles wide, that runs centrally across the inner isoseismal lines in the direction W. 34° N., that is, at right angles to the longer axes of the isoseismal lines. This band is known as the *synkinetic band*. Its boundaries are represented by the broken lines in Fig. 28. Outside this band, and throughout all the rest of the disturbed area, the interval between the two parts was one of rest and quiet. Its average duration was 3 seconds.

The isoseismal lines of the earthquake of July 3, 1904, are represented in Fig. 51. In this case, the innermost isoseismal line is a small circle with its centre close to Ashbourne. The surrounding curves are elongated in almost the same direction as those of 1903, namely, N. 31° E., and, as before, they are farther apart on the north-west, than on the south-east, side. The shock again consisted of two distinct parts, except near the boundary of the disturbed area and within a narrow central band. The boundaries of this band cannot be determined with accuracy, but its central line is indicated by the broken line in Fig. 51. This line is curved, its concavity facing the south-west, and it crosses the longer axes of the isoseismal lines at right angles and at a short distance on the north-east side of Ashbourne.

The most significant feature of these earthquakes is the double nature of the shock. A single impulse in one focus might be duplicated either by underground reflection or refraction, or by the separation of the vibrations into condensational and distortional waves. The first supposition is inadmissible, owing to the widespread observation of the double shock; the second cannot be entertained, as it would require a continual increase with distance in the interval between the parts of the shock; and, on neither supposition, can the existence of the synkinetic band be explained. For the latter reason, the conception of two impulses in a single focus is negatived. It follows, then, that both earthquakes originated in two distinct foci. One epicentre must be close to Ashbourne, near the centre of the inner isoseismal of the earthquake of 1904; the other probably about 3 miles west of Wirksworth, and the distance between them about 8 or 9 miles.

Now, if the impulses in these two foci occurred simultaneously, the vibrations from both would coalesce along a straight narrow band traversing the disturbed area midway between the two epicentres and at right angles to the line joining them. If, however, the Wirksworth focus were first in action by a second or two, the synkinetic band would be curved, its concavity facing the south-west, for the vibrations from the Wirksworth focus would travel farther than those from the Ashbourne focus before the two series coalesced. If, again, the impulse at the Wirksworth focus had preceded that at the Ashbourne focus by several or many seconds, the vibrations from the former would be felt over all the disturbed area and there would be no synkinetic band.

Thus, in the Derby earthquake of 1903, it is clear that the two impulses occurred in the two detached foci at absolutely the same instant. In 1904, the Wirksworth focus was first in action by a second or two. This slight precedence of one impulse was also manifested in the Hereford earthquake of 1896 and the Stafford earthquake of 1916. In the Doncaster earthquake of 1905 and the Swansea earthquake of 1906, there was no trace of a synkinetic band, one focus being in action several seconds before the other.

These conclusions are supported by the forms of the isacoustic lines, which have been drawn for the Hereford earthquake of 1896 and the Derby earthquakes of 1903 and 1904. Those of the Derby earthquakes are represented by the dotted curves in Figs. 28 and 51. In each earthquake, the isacoustic lines are distorted in the direction of the synkinetic band, the reason being that the sound-vibrations from the two foci were heard simultaneously along and near this band, that they were therefore louder and heard by a larger percentage of observers in this district than elsewhere.

Each of the Derby earthquakes was followed by a simple earthquake of slight intensity—that of 1903 after 40 days (on May 3), and that of 1904 after 8 hours. In each case, the isoseismal axes are approximately parallel to those of the principal earthquake, and the epicentre lies midway between the two epicentres of the principal earthquake. The after-shocks must therefore have originated along the same fault as the principal earthquakes, and in the region between the two detached foci of these earthquakes*.

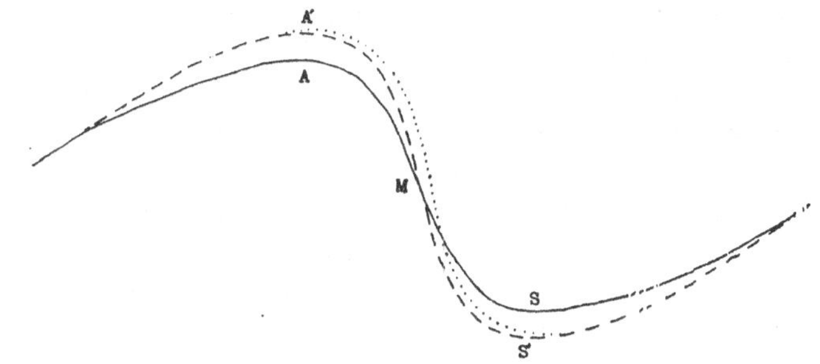

Fig. 99. Diagram illustrating the nature of the displacement that causes a twin earthquake.

243. Origin of Twin Earthquakes. It is possible that the rock along a fault-surface might be on the point of slipping in two detached but neighbouring regions, and that the waves resulting from a sudden movement in one region might precipitate a similar movement in the other. But such an explanation is inadmissible for twin earthquakes in which a synkinetic band exists; for, in such earthquakes, the impulses either occur simultaneously, or the second impulse occurs before the waves from the first focus have time to reach the second. In a simple earthquake, the displacement which produces it is probably a mere translation of the moving rock. In a twin earthquake, the only method by which practically simultaneous movements can take place in two detached foci, with little, if any, movement in the interfocal region, is one of rotation about the latter region.

* C. Davison, *Quart. Journ. Geol. Soc.*, vol. 60, 1904, pp. 215–232; vol. 61, 1905, pp. 8–17.

The continuous line in Fig. 99 is supposed to represent a section of a thin stratum in a great crust-fold along a fault at right angles to its axis; *A* the crest of the fold, *S* the trough, and *M* the median limb. Now, if a small step were to take place in the growth of the fold, from the form represented by the continuous line *AMS* to that represented by the broken line *A'MS'*, there would evidently be two regions of displacement, one between *A* and *A'*, the other between *S* and *S'*, while in the intermediate region of the median limb *M*, there would be little or no displacement. The two displaced regions would thus be the two seismic foci of the twin earthquake, more or less completely detached owing to the almost imperceptible movement of the median limb.

If this explanation be correct, the distance between the foci should be approximately the same as that between successive crests and troughs of crust-folds. Both distances are subject to wide variations. The distance between the twin foci of British earthquakes varies from 4 to 23 miles, the average distance being about 10 or 11 miles. For the lengths of British crust-folds, we have no detailed measurements, but the courses of the principal anticlines in France have been mapped, and the average distance between successive anticlines and synclines along several lines lies between 9 and 12 miles.

A movement of rotation of a crust-fold, such as that described above, must produce an increase of the stress already existing in the median limb, an increase that, sooner or later, must cause a simple displacement of the limb into a position such as that indicated by the dotted line in Fig. 99. Thus, a twin earthquake should be followed, as in the Derby earthquakes, by a simple earthquake in the region of the fault lying between the two foci of the principal earthquake.

Such after-shocks, however, are invariably slight compared with the twin earthquakes themselves. They show that the subsequent displacement of the median limb is small compared with that which takes place in the crest and trough of the fold; that, in other words, the crust-fold in its growth becomes accentuated in form more than in its advance along the surface of the fault which intersects it*.

* M. Bertrand, *Compt. Rend. Acad. Sci. Paris*, vol. 118, 1894, pp. 258–262; Davison (1), pp. 32–33.

244. Deformations of the Crust during Twin Earthquakes. Many earthquakes of great, though not of the greatest, intensity are twin earthquakes, as, for instance, the Neapolitan earthquake of 1857, the Andalusian earthquake of 1884, the Charleston earthquake of 1886, the Riviera earthquake of 1887, and the Messina earthquake of 1908. The warping of the crust during the Messina earthquake has been described in sect. 89. If Omori be correct in his interpretation of the crust-movements observed with the Formosa earthquake of 1906 (sect. 87), it would seem probable that this was also a twin earthquake. The fault along which the movement occurred was a transverse fault, but the structure of the country is somewhat different from that suggested in the preceding paragraphs. In Formosa, the fault apparently separates an anticline from a syncline in each half of the fault, the shaded areas in Fig. 41 representing the synclines, and the unshaded areas the anticlines.

Origin of Complex Earthquakes

245. The origin of complex earthquakes has already been considered in dealing with the deformations of the crust (Chapter V). The fault-displacements there described were divided into four classes, the last of which is possibly connected with twin earthquakes (sect. 87). In the others, the displacements were: (i) mainly horizontal, (ii) partly horizontal and partly vertical, and (iii) mainly vertical.

(i) The earthquakes in which the displacement is mainly horizontal are apparently connected with strike-faults. The displacement usually occurs over a great length of fault (in one case over 290 miles), both sides (in two known cases) move in opposite directions and the amount of the displacement diminishes rapidly as the distance from the fault increases.

(ii) The second class consists of the Mino-Owari earthquake of 1891. In this case, a displacement occurred along a transverse fault, which crosses most, if not the whole, of the main island of Japan, as well as along a secondary fault without visible displacement at the surface. Apparently, the whole crust in the neighbourhood of the faults was thrust forward in a south-easterly direction, the portion on the south-west side of the main fault advancing farthest. An important difference between twin earthquakes and complex earthquakes of this class should be noticed. Both are due to movements along transverse faults, but twin earthquakes are connected with the growth of a fold, complex earthquakes are due to the bodily displacement along a great length of the fault.

(iii) When the movements are chiefly vertical, they are confined to one or several vertical faults or "blatts" or to the upshoot faults of a thrust-plane. Either both sides of the fault move in opposite directions, as in the Alaskan earthquake of 1899, or one (the mountainous) side alone is moved and uplifted, as in the Wellington earthquake of 1855. In both cases, the effect of the uplift is to raise one or more large mountain-blocks and to tilt them slightly in the direction away from the fault or faults. The Assam earthquake, according to Oldham, was caused by a displacement along a thrust-plane, 200 miles in length, at least 50 miles in width, and not less than 6000 sq. miles in area—a displacement which involved minor movements along branch-faults and a general crumpling or warping of the surface-crust*.

Origin of Accessory Shocks

246. Origin of Fore-Shocks. The origin of accessory shocks is connected with the growing stresses which culminate in a great displacement, and with the various residual stresses which are brought into action by that displacement.

The stresses which end in a fault-slip may be the growth of many years, of centuries even. Neither stresses nor the resistances opposed to them can be uniform throughout the whole fault-surface. Local obstacles must be overcome before any great general movement can take place, and the removal of these obstacles is effected by local slips, each of which results in a fore-shock. The chief purpose of these slips is thus to equalise the effective resistance to motion over the fault-surface, so that ultimately, when the stresses throughout exceed the resistances, the movement takes place almost instantaneously or with great rapidity over a long expanse of the fault.

247. Origin of After-Shocks. With the displacement which gives rise to the principal earthquake, there at once ensues a

* Oldham, pp. 164–179.

change in the stresses to which the neighbouring crust is subjected. The sudden increase of stress is relieved by slips—a few considerable, the majority very small—along the fault-surface. At first, the area of this displacement extends outwards; then it leaves the terminal regions and shrinks continually towards the central regions. If the displacement of the principal earthquake be partly or mainly an uplift, the weight of the elevated mass aids its return to the position of equilibrium and must be the cause of innumerable after-slips. The after-shocks of such earthquakes should be far more numerous than those which follow a twin earthquake or an earthquake caused by a displacement that is mainly horizontal. In the Concepcion earthquake of 1835 and in the Assam earthquake of 1897, there is some reason for thinking that the multiplication of after-shocks may be closely connected with the subsequent fault-movements (sect. 216)*.

Origin of Earthquake-Sounds

248. Of the sound-phenomena which accompany earthquakes, two are of special significance with regard to the origin of the sound. These are the general precedence of the shock by the sound (sect. 73), and the excentricity of the sound-area with reference to the isoseismal lines (sect. 72).

As the vibrations which form the sound and shock travel with approximately the same velocity (sect. 73), it is evident, from the precedence of the sound, that the two sets of vibrations originate, in part at any rate, in different regions of the focus, and that the region from which the early sound-vibrations proceed lies outside the other. The excentricity of the sound-area leads to the same conclusion. It implies that the origin of the sound-vibrations lies principally in the upper and lateral margins of the seismic focus, for the vibrations from the upper margin would be more readily audible than those, if any, which come from the lower.

In the case of a fault-slip, the seismic focus is a surface inclined to the horizon. In its simplest form, there is a central region in which the relative displacement of the two rock-

* This suggestion is supported by the occurrence of the maximum epoch of the diurnal period of after-shocks (sect. 204) at about the same time as that of the diurnal period of barometric pressure, namely, about midnight.

masses is a maximum, and this is surrounded by a margin in which the relative displacement is small and gradually dies away towards the edges. As the vibrations of great range are also of long period, it is evident that, from all parts of the focus, there start together vibrations of various range and period—the large and slow vibrations from the central region, and the small and rapid vibrations chiefly from its margins. Now, between the sound-vibrations from the margins and the large vibrations from the central region, there can be no discontinuity of period. Among the vibrations must therefore be included the deepest sound that can be heard by the human ear. It is evident, also, that the intensity of the sound must gradually increase until the shock is felt, after which it must die away. Lastly, the greater strength of the vibrations from the central portion will render audible vibrations of longer period than those which come from the margins, and thus the loud explosive crashes which are heard near the epicentre should accompany the strongest perceptible vibrations.

The magnitude of the sound-area depends chiefly on the dimensions of the seismic focus and therefore of its lateral margins. That of the disturbed area depends partly on the size of the focus but chiefly on the initial intensity of the vibrations from its central portion. Thus, with very strong shocks, the sound-area may be a comparatively small district surrounding the epicentre. With slight shocks, the marginal region may be so great compared with the central portion of the focus, that the sound-area may overlap the disturbed area. In the limiting case, the central portion of the focus will disappear, and a sound will be the only result of the movement that is sensible to human beings. Thus, the earth-sounds, which are so prominent among the after-shocks of a great earthquake, are merely the representatives of creeps along the fault-surface—creeps that are so small that they do not give rise to vibrations that can be felt*.

249. Two other points may be referred to here in connexion with the origin of earthquake-sounds.

(i) The district represented in Fig. 100 is the central area of the Mino-Owari earthquake of 1891. The broken lines indicate the boundaries of the strongly-shaken area and the dotted line

* C. Davison, *Phil. Mag.*, vol. 49, 1900, pp. 66–70.

the course of the fault-scarp. During the years 1885–1892, 3014 earthquakes originated in this district, and 20 per cent. of them were accompanied by sound. The percentage, however, varies throughout the area, and the continuous curves represent this variation. The meaning of the curve marked 40 is that, if any point on the curve be regarded as the centre of a small district, then 40 per cent. of the earthquakes originating beneath it were accompanied by sound.

Now, as superficial earthquakes would have a greater chance of being heard than deep-seated earthquakes, it follows that the curves of highest percentages in Fig. 100 correspond with the

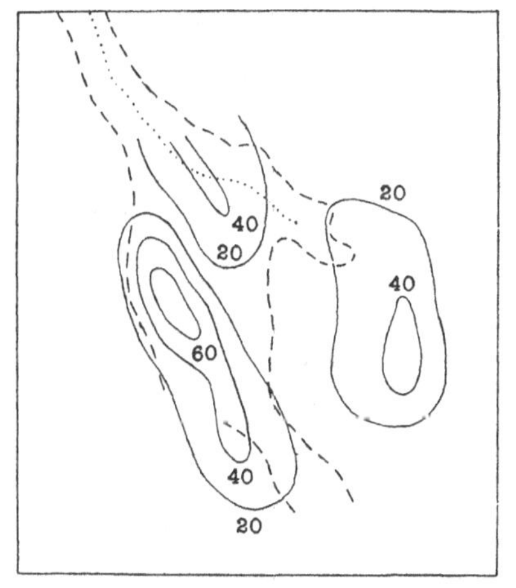

Fig. 100. Distribution of the audible after-shocks of the Mino-Owari earthquake of 1891.

earthquakes with the shallowest foci. The axes of the curves thus mark out approximately the lines of growing faults, and show that the displacement which gave rise to the fault-scarp is continued some miles farther to the south-east and that displacements also occurred along another fault following the main band of the strongly-shaken area (sects. 208, 215).

(ii) The Mino-Owari earthquake occurred on Oct. 28, 1891. In the following month, the percentage of audible earthquakes within the area represented in Fig. 100 was 18, and during the next five months it lay between 10 and 12. Then, in May 1892, it rose suddenly to 39, and during the next seven months never fell below 32, its average being 42. In certain smaller districts, the same change is noticeable. In one, the percentage of audible earthquakes rose from eight during the three months Nov. 1891–Jan. 1892 to 39 during the next eleven months; in another from 10 to 55.

Thus, the stresses produced by a fault-slip are increased in the portions of the fault adjoining the focus and especially in that above it. By slip after slip, the stresses are gradually relieved, until, even at the surface, they are no longer capable of producing the minute creeps which are perceptible to us as earth-sounds*.

CONCLUSION

250. According to the theory described in this chapter, earthquakes are merely the passing signs of the changes which are now taking place in the earth's crust. The districts in which earthquakes are most numerous and violent are those in which the crust-changes are being effected most rapidly.

In the history of the earth's crust, the period over which our seismic records extend is infinitesimal. During that period, some regions have been almost quiescent while others have been frequently shaken. It does not follow that, if the period could be sufficiently prolonged, the conditions of the two regions might not be interchanged.

So far as our evidence goes, however, the distribution of seismic activity considered in Chapter X represents the portions of the earth's crust that are now growing. There are but few parts of the globe to which the term "aseismic" can be strictly applied. Even in Great Britain, there are ancient faults which are yet in a state bordering on motion, and crust-folds that are still being intensified.

When we turn to mountainous districts of more recent growth, we find the same movements taking place, but of far greater strength and frequency. In Central Asia, some of the mountain-ranges are growing by leaps and bounds. Farther south, the Himalayan masses, as we see from the Kangra earthquake of 1905, are still being forced over the fringe of Tertiary mountains which separate them from the plains of India.

* C. Davison, *Phil. Mag.*, vol. 49, 1900, pp. 50–52.

In other districts, there are mountain-ranges in an earlier stage of growth. The western boundary of the Pacific Ocean is the most unstable region in the globe. In its festoon-shaped groups of islands, of which the Japanese Empire is typical, the crust is being pressed over the "fore-deeps" beyond just as in the older ranges of the Himalayas. In both regions, this mode of growth results in the steeply-sloping surface which is typical of our principal seismic regions.

As to the precise cause of the great and widespread movements, we are still ignorant. The cause may be one that resides entirely within the earth. But, when we consider the close coincidence of disturbances in regions so remote as the western and eastern boundaries of the Pacific Ocean, when we trace the connexion which apparently exists between the greatest earthquakes and the small migrations of the Pole, we realise that it is not impossible that we may have to look beyond our globe and recognise that other bodies of the solar system may claim a share, not only in the movements of the earth, but also in the growth of its surface-features.

INDEX

Printed in the United States
By Bookmasters